Bridging the Narrows

by Joe Gotchy

To Allan Undem

Joe Gotchy

MOSTLY BOOKS
P. O. Box 428
Gig Harbor, WA
98335

BRIDGING THE NARROWS

by
Joe Gotchy

Edited by Gladys C. Para

Published by
The Peninsula Historical Society
P.O. Box 744
Gig Harbor, WA 98335

BRIDGING THE NARROWS

Copyright 1990 by Joe Gotchy

ISBN 0-9626048-0-1

FRONT COVER PHOTO by Ken Brewer, International Panoramics Photography, P.O. Box 1932, Gig Harbor, WA 98335.

BACK COVER PHOTO by Bob Kelley, former news photographer with the *Seattle Post-Intelligencer* and with *Life* Magazine.

ENDPAPERS reproduced from "Final Report on Tacoma Narrows Bridge" by Chas. E. Andrew, principal consulting engineer, reprinted by permission.

Copyright © 1990 by Joe Gotchy. All rights reserved. No part of this book may be reproduced, stored in a retrieval system, or transcribed in any form or by any menas, electronic, mechanical, photocopying, recording, or otherwise, without the prior written permission of the author.

Copies of this book may be purchased in Tacoma Narrows area bookstores; or by ordering from: The Peninsula Historical Society P.O. Box 744 Gig Harbor, WA 98335.

Printed in the United States of America
by Valley Press
207 West Stewart Avenue, Puyallup, WA 98371

DEDICATION

This job is gratefully dedicated
to June Kline Gotchy, 1907-1985,
who was always interested in my work.
Her gifts of a scrapbook and Steinman's *Bridges*
are still helping me.

BRIDGING THE NARROWS

by
Joe Gotchy

TABLE OF CONTENTS

Dedication vii

Acknowledgments xi

Foreword by Dave James xiii

Preface xv

Chapter One: STARTING THE JOB 1

Chapter Two: HARD HAT DIVERS 17

Chapter Three: SINKING THE PIERS 23

Chapter Four: BETWEEN BRIDGES 39

Chapter Five: TOWERS 47

Chapter Six: DECK STEEL—TEN YEARS LATER 59

Chapter Seven: EPILOGUE—A TRIP TO THE TOP OF THE EAST TOWER 91

Appendix: SOME BRIDGE SPECS 97

     GLOSSARY 98

     List, TNB CREW ON DECK STEEL 100

     Letter to Joe Gotchy from Jack Hamilton 101

     TNB Layout Sections 103

ACKNOWLEDGMENTS

The Tacoma Narrows Bridge and its predecessor, Galloping Gertie, have been subjects of constant interest locally. In recent years I put together a rather inept slide show for groups asking for one, then another a bit better. It seemed to arouse the interest of even those outside the area. I began to think about a book about the bridge from my point of view after a young engineer in the audience encouraged me to share my experience more broadly.

When actually taking on such a job seemed the thing to do, I wrote to my friend, Jack Durkee, engineer with Bethlehem Steel Company. His strong encouragement inspired the first step, an important credit to him.

Others who helped this book on its way include Harold Garrett, Community Relations Coordinator with the Washington State Department of Transportation, who supplied my request for information with enthusiasm and good will.

Dave James, an old friend and retired Public Relations chief with Simpson Lumber Company, urged me to tell my story and to tell it my own way.

Ron and Pat Jones, owners of the Span Deli located off the west end of the bridge, have always been interested in the bridge itself, and their interest now generously includes my project. Ron provided the opportunity for my recent visit up the east tower by introducing me to Kip Wylie, maintenance supervisor for the TNB.

My co-workers Warren Medak, Minter, and Al Sonn, Puyallup, have been supportive and helpful with their own recollections. Bill Loomis, Tacoma, has shared his souvenir photos and pamphlets, much appreciated.

Gladys Para, who has typed and edited my material, reminded me that not all my readers would be engineers or bridge workers. And she made sure I saw all of the photos and information in the files of The Peninsula Historical Society in Gig Harbor, which include the James Bashford prints presented by the photographer's daughter Ann, Mrs. Ed Bell, Tacoma, and by the late Leonard Sund, Gig Harbor.

Bridging The Narrows

I appreciate the time and information given me by Rachel Ross Wesserling, Gig Harbor, concerning some incidents occurring during the job.

The approval and encouragement offered me by historian and author Murray Morgan has been reassuring and gratefully received.

Finally, I acknowledge the patience and helpfulness of my wife, June, who saw how important this bridge was, and the encouragement of my son, Clarence, and my grandson Thomas Gotchy, who both urged me to consider the book important too.

To all of these, I am grateful.

Joe Gotchy, Sr.
Gig Harbor, WA
1990

FOREWORD

Like the fabled phoenix, the first Tacoma Narrows bridge blew down in a November, 1940 windstorm but within a decade rose back up out of the splashes. The second bridge, twice as wide and smartly designed to let storms blow through instead of against it, stands on the original piers put in place by Joe Gotchy and his nervy fellow workers during awesome tidal sweeps 50 years ago.

Bridging the Narrows is the result of Gotchy's telling how courageous workmen daily risked their lives—and sometimes gave their lives—to spin together the fifth longest suspension span in the United Stated today.

Joe Gotchy, my friend of many years and a longtime neighbor at Rochester in Washington's Thurston County, comes from a bridge-building family. His older brother, Leonard, once the classiest baseball player in Rochester High School, worked on both the Golden Gate and San Francisco Bay bridges as well as many others. Joe likewise wound up spanning "anything that had water under it."

This is Joe's first book, written in his 86th year, about events of a half century ago. His dramatic stories and sharp pictures he had taken on the job prompted many listeners to urge him to "put it down before important history is lost."

Bridging The Narrows is the story of workmen whose names never appear on the brass dedication plates listing the governor and designer and highway officials, none of whom ever popped a molten rivet into a steel beam. Especially gripping are Joe's accounts of the divers who sank into the murky depths to position enormous anchors supporting the piers which hold up the bridge.

Joe's marvelous memory and ability to picture the processes by which only a hundred or so on-site workers strung up both the first and second bridges make *Bridging The Narrows* a unique and valuable contribution to Pacific Northwest construction history.

We live in a time when most people earn their bread working indoors, out of the wind, rain and chill. Joe's book reminds us of the vital need for, and thanks we should give for, the men

Bridging The Narrows

and women who still challenge the elements by fishing, farming, harvesting timber, building roads and bridges, putting up skyscrapers and maintaining power lines. You tell 'em, Joe.

Dave James
Author, *Grisdale: Last of the Logging Camps*

PREFACE

I was born at Bothell in 1903 and raised on a small farm at Rochester, Washington. When I, at age 19, and my brother Leonard, 24, made our start in bridge work, we found very few older men who really knew the work who were willing to share any of their wisdom. They viewed us only as a threat to their jobs. On a job with an older and experienced man, I often found that if he could find a way, he would rarely pass up the opportunity to make me look awkward.

All we had in our favor was being young and active and not afraid of heights. Experience in farm work, and the common sense it teaches, helped too. Our rigging experience was limited to what we learned by rigging up a hay fork, lifting and stowing our loads of hay in the barn with a horse as motive power. We had done a bit of logging with horses also.

Leonard and I joined the ironworkers' and pile drivers' union. Living in Seattle in 1922, we went to work for old Pop Snyder on Lake Washington. He soon made my brother a foreman. Prior to a bridge job over the Tolt River, near what is now called Carnation, we helped on the bridge over the Snohomish River at Novelty.

Jack "Rabbit Jaw" Gordon was foreman on the bridge over the Tolt. With Jack, one was supposed to be born with the ability and innate knowledge to perform the tasks required. He seemed unable to express himself in terms one understood. I finally realized, working with him, I had to look ahead and figure what the next move would be. Though difficult at first, perhaps one learns faster, working for a man like that. "Rabbit Jaw" got this moniker because he was constantly chewing tobacco.

We drove piles for the falsework across the Tolt, then rigged a boom on the piledriver and set and bolted the steel together. When it was erected we followed by replacing the bolts with rivets. None of us were riveters, so one was sent for from the Ironworkers in Seattle. A riveting gun has a piston that flies back and forth inside the gun. The snap which the piston hits on the downward stroke forms the head on the rivet, and is held in place by a spring. As long as the snap is in place the piston remains in the gun. The motive force was air furnished by a compressor.

Bridging The Narrows

The first riveter coming on the job lasted less than an hour. We were working over water and the first thing he accomplished was to drop the snap from the end of the gun, followed by the plunger, or piston. With our tools lying in the river, that ended our day. You must remember, in those days going to Seattle was not done in a few minutes.

Another riveter and more tools arrived the next day. We were just nicely organized again when this riveter dropped the snap in the river and then, looking into the barrel of the gun to see if the plunger was still there, pulled the trigger. The piston flew out and broke his jaw.

Jack had had enough. He said, "Len, you are the riveter now." I became the bucker-up who holds the tools on the end of the rivet while the gun is being used. We did very well from then on. A young engineer from Seattle, Don Evans, was our inspector. Many years later one of his relatives became the governor of Washington state—Dan Evans.

Our next job was a bridge at Sylvana, north of Everett. The last job we worked together on was the Olympic Hotel in Seattle, a steel frame building. Leonard worked on the Spokane Street bridge next, then went to California where he worked on all the major bridges including the Golden Gate. Meanwhile, the reinforced iron and concrete Beacon Hill bridge over Dearborn Street in Seattle was a learning experience for me. The ironworkers called it rat trapping—the iron, wired together for columns and other parts of the structure, did look like that.

Later, my wife June was with me on more jobs than I can name. After our youngest boy was born she spent six months in bed recovering from rheumatic fever. But a short time later she joined me in Kelso where I was working on the Longview Rainier bridge across the Columbia River. We moved, then, to Kid Valley, just 17 miles from St. Helens and Spirit Lake, where I worked on Weyerhaeuser logging bridges.

June, Joe Jr., four, and Clarence, three, and I lived in a tent one summer while I worked on a bridge crossing the Toutle River. We were in sight of the job, which made it possible for me to have a hot meal at noon. June was an excellent cook, making do with whatever was at hand, cooking on a three-burner oil stove. We were able to catch a few trout—delicious, coming from that cold water.

When the boys were six and eight, they attended four schools in one year, from Chelan to Bellingham. One of their teachers told June they were the best adjusted boys she had seen.

By the time my part of the present Narrows Bridge work was completed, the boys were both married and on their own. I drove to Puyallup every day excavating for sewer and water lines for Haeuser and Seifert. Following that job, Schuman, Johnson, Manson & Osberg hired Joe Jr. and me to go to Haines, Alaska to build a government dock. We took our wives along in May and stayed until snow flew.

I expected to be off work awhile when I came back from Alaska but it was not to be so. I was home only a few days when I was dispatched to Tacoma's St. Regis Paper Mill where they were installing a large boiler. Then Hart Construction Company had a place for me on a steam piledriver on I-5 south of Tumwater, where an overpass was being built over the Israel Road; next on an overpass at 56th Street in Tacoma; then on still another, on I-5 at Milton.

Bridging The Narrows

It was on this job I met an operator named Smith, from Seattle. He was on a diesel crane and saw I was on a steam rig. He said one day, "Joe, I'm going to give you something I have treasured for years that has been on many boilers. It was on a steamshovel I operated on the Panama Canal—when that job finished I took it with me." The next day Smith gave me a very beautiful brass steam gauge. I used it on many boilers and found it to be very accurate. On Sept. 22, 1989 I presented it to International Union of Operating Engineers Local 612 in Tacoma. I've always been sorry I never knew Smith's full name so he could have gotten credit for it.

Soon after this, in 1963, Hart Construction bought a Bucyrus crane and rigged it for piledriving and drilling. We were working on I-5 at 7th and Seneca in Seattle on Nov. 22. A parking lot attendant really shook us up when he walked over to us to say Jack Kennedy had been assassinated.

From Seattle we worked north till we finally finished our last bridge on I-5 in Blaine, near the Canadian border. I knew I would retire soon and asked to be transferred to Hart's floating steam piledriver. My last job, in 1967, was driving test piles at the old powder dock near the Nisqually Delta.

Joe Gotchy
Gig Harbor, WA
1990

Joe Gotchy, Sr., Gig Harbor, stands at the west approach to the Tacoma Narrows Bridge. The Peninsula Gateway photo

BRIDGING THE NARROWS

by
Joe Gotchy

Bridging The Narrows

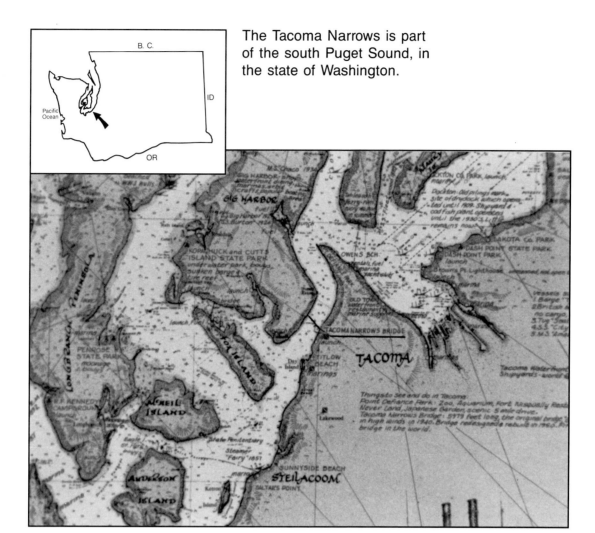

The Tacoma Narrows is part of the south Puget Sound, in the state of Washington.

The year 1990 marks the 50th anniversary of the completion and destruction of the first Tacoma Narrows Bridge, and the 40th anniversary of today's bridge of the same name.

The first Tacoma Narrows Bridge was quickly nicknamed "Galloping Gertie" for its reaction to winds, and collapsed Nov. 7, 1940, four months after opening to traffic July 1, 1940.

World War II dictated the salvage of Gertie's steel and wire, and a delay of the attempt to rebuild.

Today's Tacoma Narrows Bridge opened Oct. 14, 1950. Both bridges were, on their opening dates, the world's third longest suspension span.

Bridging The Narrows

A Foss tug brings a steam piledriver to Tacoma's 6th Ave. (Titlow Beach) dock, to begin the first step in constructing the Tacoma Narrows Bridge—building a service dock, January, 1939. A ferry approaches from Point Fosdick on Hale Passage, background. Fox Island on left. *The Peninsula Historical Society collections*

Bridging The Narrows

A Hart Construction Co. steam piledriver begins the service dock at Titlow Beach at end of Sixth Avenue, Tacoma. From left, Butch Pollock, Joe Gotchy, George (?), Bill LaChapel. A.L. Thompson Photo Service

Bridging The Narrows

The onshore service dock for the first Narrows bridge takes shape; steam piledriver on right, a rented crane on left side sets the timber decking. A.L. Thompson Photo Service

Starting the job, Jan. 2, 1939. Left, Butch Pollock, Pacific Bridge Company engineer; Joe Gotchy, right.
Bashford photo, The Peninsula Historical Society collections

Chapter One

STARTING THE JOB

I might say my crew and I received the first payroll checks for the Narrows Bridge. On Jan. 2, 1939 I was foreman on a job for Hart Construction Co. that was a preliminary to the Narrows work. We were building a dock at the foot of Sixth Ave. in Tacoma to be used as a service dock during construction of the bridge.

As work on the dock progressed, Pacific Bridge Company superintendents Jack Graham and Walt Cathey came on an inspection tour. Butch Pollock, engineer and inspector for Pacific Bridge, introduced us, and Walt turned to me and asked, "Did you have a brother on the Golden Gate Bridge, Leonard Gotchy?"

"I sure did," I replied.

My brother Leonard and I had gone to work on bridges in King County when I was 19 and he was 24. Len went on to California, where he was a very good bridge man in any capacity. Walt had thought highly of him, and when they found I was a foreman on a job that was preparation for their work, I was hired on the spot.

"When we start, you have a job as rigger foreman on the diving scow, if you want to work for us," Cathey said to me. He put me on a night shift, which had much less supervision, and the word soon got out that any good rigger (a worker experienced in moving heavy equipment) I knew could get work at Pacific Bridge.

Lyle Lewis was foreman on the steam piledriver for Hart Co. While we finished the service dock Lyle went to driving the piles for the concrete mixer dock across the water, to be situated between the west pier and west shore. There the water was deep enough to float heavily loaded gravel scows at low tide, but close enough to shore for good penetration by the piling. I had a few shifts of driving piles through its deck for the concrete mixer to set on.

Bridging The Narrows

Building the mixer dock on the west shore, where all concrete for bridge and for west anchorage was made, two piledrivers are at work here. On left is floating rig; Gotchy's skid rig on right.
A.L. Thompson Photo Service

An aerial view shows the completed mixer dock in operation between the west pier site and shore.
The Peninsula Historical Society collections

Bridging The Narrows

But meanwhile, when we were through on the service dock we were sent around to the Henry Mill on Commencement Bay to frame a portable tool and lunch room, which could quickly be bolted together after being transported to the mixer dock. I noticed other activity at the mill, which I investigated, and found that the brothers John and Don Buchanan, members of a well known lumber family, had the contract to fabricate the forms for the Narrows Bridge piers. These forms were designed for strength and much rough handling, as they were used over and over. I don't remember their ever giving us a problem during the sinking of the piers.

Shortly after this I went to work for Pacific Bridge, while they built the west and east piers of the Tacoma Narrows Bridge. Most of my crew were glad to go with me. We began unloading supplies onto the service dock as they arrived in a steady flow. Besides the spools of cable— thousands of feet of various diameter—there were a number of two- and three-drum winches that had been steam powered but now were converted to run on compressed air. We unloaded a compressor, a splicing shed to hold the cable, a tool room and an office.

Besides everything else on the service dock there was a decompression tank large enough to hold two bunks and an electric heater for use by the divers, if necessary. The entrance had a small compartment called an airlock that allowed a doctor to enter if his services were needed. A diver with the bends is a very sick man till he can be put under compression. The heater was a necessity, I found. After we had the tank hooked up to air I tried it out. The valves were so arranged that the air pressure could be controlled from inside or outside. With the doors closed and the pressure building, the air became quite heated; but when I drained the pressure off, it fogged up so densely I could not see 12 feet and the chill was sudden, and very cold. I was told later that when the divers used the decompression tank they warmed themselves up with hot coffee royal, made on a hotplate.

From that well equipped service dock a man-scow, towed by one of the many tugboats on the job, carried me and my crew to work, dropping us either on the diving scow or one of the rigging scows moored at the pier. If there was diving to be done that day, the diver came on the scene whenever the tide dictated, and after we had rigged the diving scow for the job.

The man-scow transported three crews daily to their jobs on the west and east piers. The **Louise** provides the power here.
The Peninsula Historical Society collections

Bridging The Narrows

Bosses on the Big Job is the title given this photo offered readers by *The Tacoma News Tribune.* From left, Ralph Keenan, general manager, and Walt Cathey and Jack Graham, superintendents, Pacific Bridge Company. *Tacoma News Tribune* photo

My first job on the diving scow was a tour we made of the bottom before starting the west pier. A diver was lowered riding a bosun chair just three or four feet above the bottom, while a boat moved us slowly at slack tide. The diver was Johnny Bacon, a small man but one of the best. He relayed over his phone to his tender what he was seeing in 120 feet of amazingly clear water, even his view of little octopus.

Jack Graham, Walt Cathey and Ralph Keenan, the general manager for Pacific Bridge, were on this tour. They agreed that the west pier could be started with no site preparation.

The 550-ton reinforced concrete pier anchors were being cast at Scofield's, located at the old 15th Street bridge in Tacoma. They measured 12 ft. x 12 ft. x 51 ft.-6 inches, and were towed to the Narrows one at a time, to be dumped in a 900-foot circle around each pier site with no lines. That meant two or more dives on each of the anchors for each pier, first to attach temporary lines to a rigging scow, and later the cables that would connect anchor to pier. On the first dive, the two rigging scows, anchored side by side over the exact pier site, received the temporary cables from each anchor which were made fast. With so many cables to each scow, tags were necessary.

Bridging The Narrows

Onsite, this group is identified as follows, from left: Jack Graham, Walt Cathey of Pacific Bridge; Fred Dunham, resident state engineer; Orville Sund, general manager, Foss Tug; Butch Pollock of Pacific Bridge; Heffernan, manager, Glacier Sand & Gravel; Davis. The Peninsula Historical Society collections

Pacific Bridge superintendents Walt Cathey and Jack Graham, and Harvey Donnelley, chief inspector for the State of Washington. Bashford photo, The Peninsula Historical Society collections

Bridging The Narrows

Forms encase two concrete pier anchors, each poured in place on a Foss barge at the Scofield plant. Tacoma's copper-domed Union Station appears center background. Bashford photo, The Peninsula Historical Society collections

Bridging The Narrows

I should state here that Johnny Wyckoff, state engineer, was the instrument man who spotted the anchors. Then, when we had to dive on them, he did a very accurate job lining us up. Not long after the bridge was finished he became hull engineer in the shipyard for Todd; later he was pit manager for Glacier Sand and Gravel.

The first one of the west pier's 24 anchors was towed onsite Jan. 24, 1939. The scow carrying it was spotted in the water as over a target by signals from shore, then its sea-valves were opened to flood the scow to a tilt. The anchor slid off, causing an upheaval ending with a terrific flop of the emptied scow. I am told that because of it, the caulking had to be gone over each time a scow delivered an anchor.

The Foss Company supplied the bridge job with several of their tugs and scows. I remember hearing a conversation among Pacific Bridge officials concerning the Foss contract. At the time, Foss crews working in Puget Sound worked a 12-hour day. Pacific Bridge decided, "No, an eight-hour day will prevail. These men are working on a hazardous job and we want them alert at all times."

One anchor was lost while rounding Point Defiance. It is a wonder more of them were not misplaced on the Narrows site, with its rushing tides. To my knowledge only one anchor gave us trouble, because of its positioning, while connecting the 1-9/16-inch cable to the pier, though that job was always a struggle. The fast-flowing tide could put enough deflection in the large cable to make it seem shorter than it actually was.

The pier anchor is cast aboard its carrier on heavy timbering that enables it to slide off after its trip to the Narrows site.
Bashford photo, The Peninsula Historical Society collections

Bridging The Narrows

The *Tango* tows a concrete anchor to the west pier site, Feb. 4, 1939, positioning it in response to signals from shore. The *Louise* assisted, then moved out before the dump. Bashford photo, The Peninsula Historical Society collections

Bridging The Narrows

A Foss tug has maneuvered a pier anchor into position before it is dumped. Bashford photo, The Peninsula Historical Society collections

Dumping the 550-ton anchor is accomplished by opening sea valves to tip the scow. A.L. Thompson Photo Service

Bridging The Narrows

A pre-formed caisson for one of the two main piers is built at Pacific Car & Foundry Co., Seattle, The steel cutting edge appears on lower portion; sheathing is being applied to the structural steel frame before concrete cells are formed.
Bashford photo, The Peninsula Historical Society collections

With anchors in place the piers could be started. One of the last steam tugs Foss owned, the *Wanderer,* towed the west pier caisson, afloat, from the Seattle area where it was fabricated. A huge structure of hollow concrete cells sheathed in wood, with a bottom finished by a steel cutting edge, it rose out of the water for at least 50 feet, its flat top surface measuring roughly 60x120 feet. It arrived too late in the day to be put in place and had to be moored at the service dock overnight. Several of us riggers were put aboard the caisson to help in the mooring. It floated very high above the dock, which was crowded with people all with upturned faces, and I remember wondering how we could throw those weighted casting lines to the dock without hitting someone.

The next day the two rigging scows at the pier site were separated and the pier was floated between them. L-shaped fenders that we had previously built at the service dock were floated into place, so the pier was completely surrounded with an independent float of 12x12s decked over with planking, plenty of stanchions to moor anything to, and all well bolted together. This caisson was to become the bottom section of the finished underwater pier, gradually sinking while guided by the anchor cables, as each new section was constructed on it above the water. Its cutting edge would, upon completion of the pier, bite into the compacted sand and gravel on the bottom of the Puget Sound Narrows.

Bridging The Narrows

The caisson for the east pier was towed to the site in August, 1939 by the 128-ft. *Wanderer*, the Foss Company's last steam tug, built in 1890 at Port Blakely. Bashford photo, The Peninsula Historical Society collections

Bridging The Narrows

There never was a problem in pier flotation as they were being constructed, but I do remember that two pumps were installed to remove any possible leakage in the temporary wooden bottom of the caisson. The piers remain to this day quite hollow, with a concrete seal at the bottom, and at the top under the pedestal on which the towers are set.

Now we began replacing the temporary lines to the concrete anchors with rust resistant cable of 1-9/16-inch diameter and a breaking strain of 175 tons. The concrete anchors and their heavy lines performed the same function as the guy lines to the spar tree from surrounding stumps did in high lead logging. When we finished, the 24 sets of rigging holding the west pier in place all through its construction had lines adjusting the tension which were clamped off on the pier. I was told much later that when the bridge was completed the anchors' connecting cables were simply cut away from the piers and dropped, and not retrieved. Fishermen whose lines become fouled in the area may not be surprised to learn that.

If my memory serves correctly, Star Iron & Steel, Tacoma, fabricated the four guy derricks that worked two to each pier. They were designed with the expectation that as the pier was sunk with the added steel and concrete, one derrick would lift the other to the new level and, in turn, be lifted into place. When the first pour on the west pier was finished, the derricks were rigged to move up to the next level. For some reason the first time this was attempted both derricks collapsed in a tangle of twisted metal.

I was watching the activity from the Sixth Ave. dock, easily seen in the clear, no-fog conditions that evening, just coming on shift. The sudden derrick collapse was something we had not expected. Looked like serious injuries to us and it was, for two men. A worker named Corcoran had his thigh punctured by a reinforcing rod, and my friend Al Sonn received a concussion and severe gash in his scalp. He felt lucky, he said, to get off so lightly. Electric winches replaced the lifting derrick, after that incident.

The pier was sunk by adding sections of structural steel and concrete cells. Its derricks, called "creeper" or "jumper" derricks, were designed to raise each other alternately to the next level as the piers sank with each addition. The first time this was tried a boom on the lifting derrick collapsed, dropping the one being lifted, shown here.
A.L. Thompson Photo Service

Bridging The Narrows

Following the derrick accident, electric winches were used for lifting. The jump has been completed using one at each corner of the two derricks. Detail of pier sections seen here are a) the hardened concrete of a previous pour at waterline; b) wood forms to contain the present pour; c) structural iron frame of section to be poured above it. The Peninsula Historical Society collections

A derrick lift has just been completed, using four electric winches to lift each derrick up to the next level of construction. The pier sinks slowly back to waterline with the weight of each completed section. A rigging scow, cluttered with necessary equipment, holds snug against the pier in foreground. A.L. Thompson Photo Service

Dredging for the east pier began April, 1939 in a strong-running tide. West pier construction and the west shore concrete mixing plant appear in background. Bashford/Thompson photo, The Peninsula Historical Society collections

Bridging The Narrows

Building the east pier was a very different operation. The slope there was so great a dredge worked for quite awhile before a flat area large enough for the pier was achieved.

The method here was unusual. A rigging scow, with a compressor, air winches and other equipment was anchored some distance west of the pier site. The water flowed so fast several anchors were needed to hold the scow in place. The dredge had its problems holding its position, so a line from a winch on the rigging scow was used on the port side for holding and shifting. At a given signal from the steam whistle on the dredge, the operator on the rigging scow was to give or take in slack.

I felt sorry for that man. He was alone much of the time, in the dark and deathly afraid of that fast-running water. We had to spend part of my 4:30 to midnight shift on that rigging scow. When my work was finished I walked around the deck and inspected all lines. As I came to the forward port stanchion I could see it was pulled away two inches; if it moved more, a serious leak could occur. One of the heavy anchor lines was clamped off here, so I put a two-part purchase on it with a winch line and pulled it back in place, then dogged the drum so it would stay. I told the operator under no circumstances to release it.

The next day when I came on shift I learned the entire side had pulled off and the scow was sunk. The boat crew told me they picked the operator off just as the scow went down. A lot of equipment was lost, very little recovered. Divers went down and a Smith clamp, also called a come-along, was recovered, its value about $400. Winches, air compressor, light plant and scads of tools were likely scattered from there to Point Defiance, the way the tide was running.

When conditions were studied afterward it seemed the only thing that could have happened was that the operator got a whistle for slack, released the line I had put on instead, and, as the dredge moved away it pulled the rigging scow away from its securely anchored side. Later I heard the Pacific Bridge officials discussing this. John Blondin, a chief engineer, remarked, "It may be a good thing this happened. Now we will have enough respect for this fast tide action to avoid a more serious accident."

Working on a job of this magnitude was a constant challenge. As I look back on it now, though, the planning and coordination were excellent. Everything seemed to fall in place without anyone getting overly excited about the few setbacks that occurred. The men of the Pacific Bridge Company had the Golden Gate Bridge and the Bonneville Dam and Locks experience behind them. The teamwork displayed by their entire organization, from the manager on down the line to their foremen, was exceptional.

Jealousy between foremen is something I had known from the past. There was none of that on this job. Clate Rand, the foreman who preceded me on the day shift, always did his best to leave me in a very comfortable position. If anything had changed he explained to me in detail. Rand and superintendent Kitchen, and two brothers, Leon and Bill Dippold—both about six-foot-four, hard workers and always full of fun—were all from the Bonneville job. When the Narrows job was over the Dippolds went to Pearl Harbor just in time to be caught in the hostilities. Our union received a letter from them to let us know they had survived the Dec. 7, 1941 surprise.

Bridging The Narrows

John Blondin, a chief engineer with Pacific Bridge, enjoyed diving while on pier-inspection duty. A.L. Thompson Photo Service

Chapter Two

HARD HAT DIVERS

The divers employed by Pacific Bridge were Johnny Bacon, Bill Lahti and Chris Hansen. Bill Reed was a diver employed by the state, and John Blondin often enjoyed diving while carrying out his duties inspecting the piers. The work these divers could accomplish in the dark was amazing. Diving time on the bottom was short. On the west pier, depending on the tide, their working time ran from 40 minutes to an hour and 20 minutes, and usually less than 60 minutes, so everything was prepared carefully to expedite the actual dive.

The diving scow was equipped with a light plant, compressor, a small decompression tank and plenty of gear, all inside a snug house adequately heated by the compressor. Outside it were air winches, I think three, with an air tugger on an A-frame at the very front of the scow, on which the divers rode up and down. Working at night to change lines on the concrete anchors, we used a manilla line 3/8th inch by 100 feet, fastened to the weighted bosun chair that carried the diver. If he failed to land on target, he would take the finder line and make a circle until he found what he was looking for. Then, when ready to come back to the surface all he had to do was follow the finder line to his private elevator.

There is something I'll mention here that I doubt anyone today besides me is aware of. The air from those compressors was tainted with oil fumes, which had to be removed. All the air the divers breathed went through a small canister packed with Kotex. When I first heard of this I thought it was a joke. But the divers' tenders had to change this regularly.

Two of the pier anchors had broken when they hit bottom. On my shift we had to wrap one with cable, a difficult feat for the divers. Fortunately, the divers were able to do this in daylight, working as a pair, being careful not to tangle their lifelines, telephone lines and air hoses. First, we passed down stirrups that fitted over the top and hung down the sides of the anchor about halfway, upturned to lay in the cable. Then, moving with caution, we passed cable and clamps down to them. A lot of work for one tide, but they did it. Johnny Bacon and Bill Lahti worked well together.

Bridging The Narrows

On the diving scow: Bill Reed, diver; his tender, kneeling; Johnny Bacon, diver; Roy, his tender; Johnny Wyckoff, state engineer. The Peninsula Historical Society collections

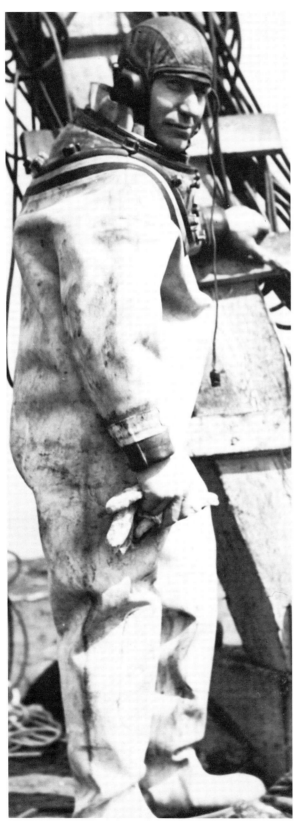

Bill Lahti, diver. Bashford photo, The Peninsula Historical Society collections

A diver surfaces after a successful dive.
The Peninsula Historical Society collections

Bridging The Narrows

Bacon and his tender, Roy, were small men, but an excellent diving team. Of all the tenders working with their divers on the job, Roy impressed me the most by the attention he gave to Johnny's safety while diving. I never knew of either having an alcoholic drink at any time. Another diver, with his son as his tender, made a totally different team. While he was a very experienced and capable diver, he seemed always to reek of alcohol while I was helping to suit him up for a dive. All the divers working with us had solid past experience in their risky, difficult jobs.

The diving equipment used by these "hard hat" divers was much different from what is worn by Jacques Cousteau's men on TV. A hard hat diver was dressed in a heavy, rubberized suit, a metal—usually brass—helmet having a glass faceplate, and lead shoes. This created a uniform that, with him inside, weighed about 384 pounds. The shoes weighed perhaps 20 pounds each, making it awkward to move around out of the water. Underwater, with his suit inflated to displace 6.5 cubic feet, he weighed less than the water he displaced. The sea water weighs 64 pounds per cubic foot, or 416 pounds, and thus becomes a force buoying him up. To stay on the bottom and perform any work he must add even more lead to his bulk, in the form of small weights hung from a belt supported by shoulder straps, as many as he thought necessary.

A steel frame sitting on the diver's shoulders carried the helmet. After the helmet was screwed into place an airtight seal was formed by tightening special nuts onto studs projecting through the vulcanized suit, locking the helmet and suit into one unit. The air hose, lifeline and telephone cable were solidly connected to the helmet. Lastly, of course, all divers on the bridge job wore a name tag with a notice reading, "If Found Unconscious Rush to the Foot of Sixth Ave," where they would have access to resuscitation. That tag saved a couple of men from arrest for drunkenness.

Bill Lahti's tender was late coming to work one time. He asked me to help him dress so as not to lose the tide. The final straw was to tighten the nuts that required a special wrench, the wrench the absent tender had with him. I was finally able to do the job with a crescent wrench. Just before he closed his face plate Bill said, "This is the first time a rigger dressed me with a crescent wrench." His tender came, and we got on with our work.

Bill was a native of Finland, and had served time as a sailor on the old windjammers. He had a lot of stories saved up from that period of his life. The food usually was very poor aboard ship, he told me, and one captain kept a pen of chickens for the eggs and an occasional chicken dinner. But Bill noticed that when a chicken died naturally the crew had chicken dinner, too. Very shortly he came up with a bright idea. He made a small stick long enough to go through the wire fence, with a needle embedded in the end. When he figured it was time to eat chicken he very slyly poked the stick through the fence, and enticed the chickens near with a few bread crumbs. Then, lining his needle on the head of a fat hen, he gave it a quick jab and the crew soon had chicken dinner. No one wised up as to what disease caused the death of those chickens.

Though Chris Hansen was not on many of the dives on my shift, the foreman on another shift told me about the time the tide changed while Chris was down on an anchor and refused to come up. His time on the bottom was fast vanishing. They tried to pull him up by his lifeline, but he had taken a turn on the anchor line and they could not move him. Finally, they saw his lifeline go loose, apparently drifting with the tide. About 130 feet away Chris popped to the surface before their eyes, spreadeagled. He had closed the exhaust valve on his helmet so that he had flotation. Coming up that fast with air at better than 60 pounds pressure, he was lucky his suit did not burst. He had to go into the decompression tank on the diving scow. I was glad it had not happened on my shift.

A young longshoreman from Portland had asked Pacific Bridge for a test dive; my shift was chosen for his tryout. He had done some diving to 30 feet, but 120 feet is another matter. It was decided that Roy should tend him, as a safety factor. This young man was making his dive in total darkness which in itself would be unnerving, let alone the depth he had to descend to. Roy felt he was trying to work with too little air, and was kept quite busy trying to calm him down. The correct pressure gives the diver a few inches of room inside the suit; otherwise the water presses in on the diver. Had it not been for Roy's calming voice I am sure he would have panicked totally.

He may have been thinking about the particular thing that can happen, that all inexperienced divers fear, the water-hammer. If his suit should rupture and lose the few inches of inside air, the terrific pressure of the water would most surely crush his body to a pulp and try to crowd it into the rigid helmet. The old tale divers tell of such a tragedy ends with, "And they buried the helmet."

Before the diving had started I was working a split shift, sometimes with only five hours off. If I drove home 20 miles it seemed like I hardly got to sleep till it was time to get back to work. I solved that by putting a trailer back of the Hollywood Inn, just a few steps from the Narrows Bridge service dock. The home-cooked meals I could get there were outstanding, and Mrs. Sofie Wesserling made everyone feel at home.

I am indebted to Rachel Ross Wesserling, a Gig Harbor resident whom I had not seen in 40 years until recently, for the following facts. After her husband's death in 1937, Sofie Wesserling sold her business at Allyn and bought the Hollywood Inn at Titlow Beach, next to the office and shop of the Pacific Bridge Co. She was nicely settled and ready for customers when the bridge work started in 1939. The workers beat a wide path to her door. Rachel said Sofie catered to these special people, many from other states. Her welcome was warmly accepted and never abused. No unacceptable language was ever heard at Sofie's table. After hours the men were encouraged to talk about their families and home. A piano and accordion were in the place, to be played and occasionally danced to. It was a home away from home to many of the bridge workers. Mrs. Wesserling had only two employees besides herself—her son Elmer and Rachel Ross.

While I was eating there, several times I saw various pieces of equipment go by with the name BILL REED SUBMARINE DIVER in large letters on a background of bright aluminum paint. I wondered who Bill Reed was, and why he had to point out that he was a submarine diver. We had no sky-divers at that time, so what could he dive in but water? Later I realized it was an advertising ploy he had dreamed up. Rumor had it he was an ex-Navy diver.

Most of the divers I knew did not live happy ever after. When he first started on the Narrows job Bill Lahti told me that if all three original divers finished alive he'd be surprised. He said the fast tides made it a treacherous undertaking.

Bridging The Narrows

Two rigging scows lie side by side with the diving scow tied up against them at a right angle, as the tide flows past. *Bashford photo, The Peninsula Historical Society collections*

Bridging The Narrows

Within a year after the bridge was finished Chris Hansen was paralyzed, unable to work again. And Bill Reed, I heard, had a cerebral hemorrhage which ended his diving. We all felt sorry for Bill. He had turned all his pay over to his wife and thought they were well fixed financially. But it turned out she had been playing the horses with no winners.

I never did hear again from Bill Lahti. But word came back that Johnny Bacon, while diving later at Pearl Harbor, had his suit rupture and was crushed to death by the terrific waterhammer. The old tale told by divers made news once again: the pressure jammed his body into his metal helmet. And they buried the helmet.

A runaway log raft jams against the west pier. Rigging scow is caught between logs and pier pedestal. Tower appears in background. The Peninsula Historical Society collections

Chapter Three

SINKING THE PIERS

Sinking the piers was a 24-hour, seven-days-a-week job. I had an eight-man crew but each man had one day off a week, so the crew changed every day—always the same men, working different days. When the concrete pour started on the west pier things were hectic till it settled into a system. Even then, unusual things happened. Like the night a log raft on its way to Tacoma came too close in the dark, swung crosswise and broke up on the pier. We had some rigging to put in order after that, but the anchors held the pier in firm. Fortunately, the diving scow had been positioned on the other side of the pier, away from the impact of the logs.

Just hauling the concrete from the mixing dock to the pier could get hectic. At the mixer dock, pours were made into four huge buckets, each positioned in a pocket built up on the deck of the "pot scow." A fifth pocket was always left empty, to receive the last bucket left behind on the pier from the previous run. The pot scow was then towed over to the pier site by one of the many tugs working to keep up a steady delivery of concrete. The tide often made that job hazardous, as well as everything else we did at water level. The Foss skippers worked with us all through the days and nights of sinking Gertie's piers. I remember them with the highest regard—Hank Schoen, Harry Manley, Oscar Rolstad, Bruce Palmer, Leonard Sund, Vern Wright and Bill Case, the dispatcher. However, a tug named *Tango,* privately owned, also was on the job.

She was powerful, sat low in the water and was all spit and polish. The owner and skipper sure would scream if any paint was scratched. Doug is all the name I ever heard him called. No, I take that back—there were other names I won't repeat. There was no doubt Doug had a fine craft in the *Tango,* and he handled it like a teenage kid with a powerful car to play with. But his control of his tug did not always extend to his load. Doug decided to show up the Foss skippers and be the first tug to haul two loaded concrete scows in tandem to the east pier.

Bridging The Narrows

Henrietta tows her loaded "pot scow" to a pier site. Towers in background carry electricity from the Cushman Dam across the Narrows to Tacoma. Bashford photo, The Peninsula Historical Society collections

Tango attempts a double tow, two scows loaded with concrete buckets, in tandem.
Bashford photo, The Peninsula Historical Society collections

A loaded pot scow, four full buckets and front pocket empty, is towed from the concrete mixer dock to site of pier construction.
Bashford photo, The Peninsula Historical Society collections

Bridging The Narrows

The first time he tried that, I heard someone tell him, "You staggered like a drunken sailor." Doug just laughed and answered, "Well, we got here just the same." The tide was running a bit faster on the next trip, and just before he arrived at the pier the rear scow rolled over and lost four buckets of concrete. A diver went down for them later but found only one bucket, stopped against a large rock on its way down.

This tide action was a hazard we worked with at all times. Flotation vests were apparel we all wore. Even then, in that water, with whirlpools surrounding the piers, a man overboard could be pulled down. Safety precautions were closely observed but we were lucky, also.

We had been diving for several weeks connecting the heavy anchor lines to the fall blocks, adjusting them to hold the west pier in place, and getting along very well. A new diving superintendent, Jimmy McCloud, came with us on a couple of dives, though he divided his time with the other two shifts also. We had just come on shift and were aboard the tug *Louise* on our way to the diving scow. As the boat approached the scow, heading into a 10-knot current, the skipper swung the fantail stern in close enough for Marion Christian and me to leap board. Our feet had barely touched the deck of the scow when its headline anchor began to drag, allowing the diving scow to turn broadside to the current. The rest of the crew and Jimmy McCloud were aboard the tug watching us. Jimmy was a man who never ordered; he advised. When he commented, "Leave her, Guys. She's going to turn over," we knew he was telling us to save ourselves.

The scow had tilted enough so water was at the edge of the deck. Marion and I could see that the line from the concrete anchor, about to be fastened to the pier, was firmly shackled to one of our winch lines, and knew instantly we had a chance to save the scow. The deck of a working scow is necessarily cluttered with the lines from the fairleads, about a foot high, to the winches, about three feet above the deck. I said to him, "Slack all winch lines to our regular anchors!" And we worked like twins, flying over obstacles to the levers from one winch to another till all four anchor lines were slacked.

It was hard to believe how quickly that scow straightened around, hanging on the concrete anchor. Marion and I knew we were very lucky to have avoided a cold swim, if nothing worse. No doubt it would have turned over otherwise, and that would have been a serious loss at the time. My time was so well occupied on the job I never gave this incident another thought, but noticed that Jimmy did not spend much time with us afterward.

After this experience we put a 5,000-pound steel anchor on our headline. Even then, with a stern anchor and two side anchors, our scow sometimes dragged anchor, requiring the *Louise* to hold us against the rushing tide for as much as one and a half hours. We finally put 60 feet of heavy ship's anchor chain on, which held us, but made picking the anchor to move much more difficult.

Another incident occurred at the bridge pier. The *Louise* was a deep-hulled, rather narrow ex-fishing boat, about 60 feet long, with a high pilot house that gave very good visibility to the skipper. He had come in on slack tide and moored at the end of the fender. As the tide changed the boat drifted sideways, laying up tight. The tide was a fast one in a very short while. When the *Louise* was wanted to run someone over to the service dock, her skipper tried to leave but was helpless, his boat totally pinned to the fender by the fast-moving water. He had to call in the *Tango* to pull his boat free.

Construction was ongoing at both east and west pier sites with support from scows, tugs and small craft. Tacoma's unpopulated east shore is background in this view from the west shore. Bashford photo, The Peninsula Historical Society collections

Bridging The Narrows

This work was very unforgiving. A mistake at night could easily cost someone injury or his life. When the men assembled at Sixth Ave. to be transported to the job, I always cast a critical eye on my small crew. I was concerned only with the men I was responsible for. A husky young fellow, who was performing totally out of character, caught my attention one evening. It was plain to see he was drunk. He was much larger than I and he loved to fight. As we neared the pier I approached him and said, "Ernie, don't get off. Go back home. If you come out tomorrow, be sober. I'll put up with no drinking here." He surprised me by politely agreeing.

The closest we came to injury on my shift was the time we were handling the block and tackle rigging for one of the west pier's 24 anchors, never easy and always a struggle. All but one of the concrete anchors dumped around the pier had come to rest within its limits. Fred Mickelson, a good rigger and longtime friend, and I had pulled the fall block from that anchor onto the scow and were attempting to pin it to the socket on its heavy line, just a bit too far away. We could come only within an inch of getting the pin in. A no-go situation.

We were lying down reaching to the socket, about 15 inches long and on a 45-degree angle, trying to place the pin. We were placing a terrific strain on our winch line, but could not make the hookup. Fred had just moved his head up a few inches when the socket made a flashing turn that would have decapitated him if he hadn't moved that split second before. I looked at him and said, "That was a close one." He just nodded, but some of the color had left his face. We had to make up a short strap of heavy cable to replace that lost inch.

The diving scow was involved in a dangerous incident that had nothing to with our crews. It happened on the day shift shortly before I came on. The tide was running fast and two fishermen had run out of gas directly in front of the diving scow. They had used poor judgment to be in that position. With the tide running 10 knots and their boat drifting rapidly into the scow, they froze. The deck of the scow was no more than 30 inches above the water. All they had to do was step, jump or fall on their bellies onto the deck; men there were ready to grab them. But they made no attempt to save themselves. Their boat rolled under the scow.

The crew moved to the back end and waited to rescue them when they appeared, accomplishing it without incident. The fishermen were not injured, only wet and still scared. Their boat, borrowed from its owner, Al Winterhouse, had some broken ribs. Russ Schmidt, a friend of Winterhouse, had borrowed it for the day's fishing. His companion was a relative but I never learned his name. I will bet that trip under the diving scow was well remembered and maybe even caused some nightmares. The rigging gang was very happy the rescue was so successful.

When the derricks were operating and we were not diving, I usually had a list of things to do, the highest priority first. I felt responsible for my crew, who at times would be split up and working in three places. On one of those nights one of my men came to me after dark and said, "Joe, I am afraid Fred will commit suicide. His wife just left him and he is in a terrible depression." I looked up Fred on the job and we had a long talk. I told him I would be much happier if he would let me have the *Louise* set him ashore. I let the skipper know what the problem was, and he had a hot mug of coffee poured when Fred got aboard. He also engaged him in conversation to distract him from his personal problems.

Bridging The Narrows

A year later I got a phone call from Fred asking me to meet him downtown. He could not thank me enough for that night. He had landed a job as superintendent for a large contractor in Oregon, and told me any time I wanted one there was a job waiting for me. Just that he had looked me up was thanks enough.

The 24 sets of rigging that held the west pier in place all through its construction had lines adjusting the tension which were clamped off at the pier. Walt Cathey came to me and said, "Joe, I want a man to handle the tension lines. He will have to use a short pole to tap them for tension and make adjustments."

I had a man who just answered that description. Ernie Warfield had handled lines on floating piledrivers for years. I thought I was getting Ernie a good, easy job, and it proved to be a fine job for him. Even the carpenter superintendent Mac McPeak said that Ernie was the best, and that he saw at least two shifts on the job so he knew. But a fly can always get into the ointment. Ernie got to thinking about all those fat ling cod on the bottom, just waiting for a hook, and he could not resist as he had plenty of time.

It so happened that Walt Cathey came by just as Ernie was landing a big ling cod, and Ernie was terminated on the spot. Later Mac cold me it was a mistake to can him, that no replacement did as well. Shortly after, we had the diving scow over an anchor at the east pier site, waiting for slack water to dive. I saw a skiff approaching with two men. Seemed to be heading for us, so I stood by to take their line. It was in the evening, and here were Jack Graham and Walt Cathey, who was just winding in his salmon gear. I looked at Walt and said, "I'd be a little careful about fishing, Walt, if I were you. I understand they do can people for fishing on this job." Walt looked at me with a sheepish grin and said nothing. I can only say it was a pleasure to work with those two. I never had an unpleasant experience, and if I needed anything they were most cooperative.

Sinking the piers, which meant building them section by section upon the original caisson, was a continuous operation: Pour concrete, put up steel, raise forms, do it over again. The hollow cells formed from concrete reached to the very bottom, which was made of timber. The bottom edges of the lowest section were the steel cutting edges of the caisson. When the pier, floating upright during construction, reached a predetermined point in sinking, the blocks and lines that went to the concrete anchors had to be moved up to new positions by the divers. This was always done on the day shift, as good visibility helped.

Constructed to their full size onsite, the piers' exterior walls are three feet-three inches thick; the interior is divided by crosswalls of reinforced concrete two feet thick, covering the structural steel framework extending from top to bottom. Each time a pour was made the pier sank lower in the water, its buoyancy remaining the same with the enclosure of more watertight cell area. The final excavation was carried out through the hollow concrete cells thus formed.

When the floating pier was near bottom, setting it down had to wait for the right tide conditions. The anchor lines were manipulated so the pier was on the selected spot. Two heavy shafts, one on each derrick, were lowered down through the cells until they reached the bottom timbers, where they knocked the bottom into splinters. It fell to me to take off the shafting and reeve up the three-yard clamshells. These clamshell buckets were totally different from any I ever had experience with. The closing halves were levered in such a way there was

Bridging The Narrows

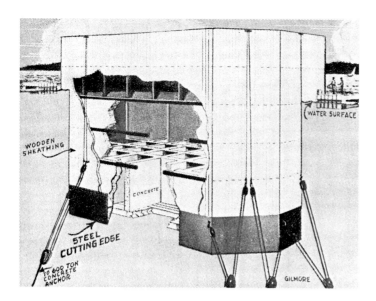

A cross-section drawing shows a floating caisson, the initial and lowest portion of the bridge pier. The pattern of its interior construction can be seen above the steel cutting edge that will be embedded below ground. Single lines on separate blocks (contrary to this interpretation) connected the pier under construction to the concrete anchors, holding it upright against the tide.

From: "Final Report on Tacoma Narrows Bridge," Chas. E. Andrew, principal consulting engineer, June, 1952

Riggers adjust lines from pier to concrete anchors on bottom: While a derrick holds tension, a worker tests it by feeling the line, giving appropriate orders to the men using wrenches. Bashford photo, The Peninsula Historical Society collections

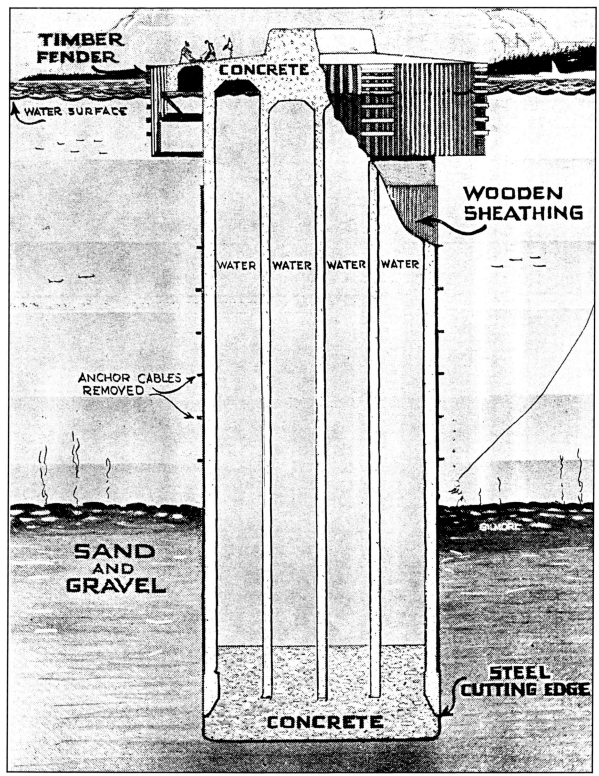

Piers were constructed of steel and concrete, with heavy timber fenders at water level. This drawing by Gilmore shows the 20-foot concrete seal, top and bottom, and the pedestal atop to receive the tower. From: Chas. E. Andrew report

Bridging The Narrows

very little lift till the bucket was closed. As I was figuring this out the master mechanic for Pacific Bridge came over to offer his help, saying, "Did you ever reeve up one of these before?" I said, "Hell, no! I never saw one before!"

While the clamshell dug up wood, and sand and gravel from the bottom of the Narrows, frequent checks were made. The pier could be steered within reasonable limits by the digging. If the pier was moving off the location, a bit more digging on the correct side would bring it back. Standing on the pier as it moved gradually down was like standing in a small earthquake. One day I spied a heavy duty claw hammer coming up in the clamshell. I have it to this day, likely the only hammer to be lost in the Narrows bridge and claimed again.

Digging for the west pier penetrated 55 feet. The east pier was dug down 90 feet, for the bottom was so sloping there an extra margin of safety was desired. Both stopped in compacted sand and gravel. Then the bottom seal was poured, using a long steel tube known as a treme. It reached within two to four feet of the bottom and had a funnel at top where concrete was poured in. A large bundle of burlap bags was crowded into the entrance of the tube before the pour. It acted like a piston, pushing the water out at the bottom ahead of the concrete, so there was no separation. The air pressure in the tube must have reached 100 pounds per square inch; at times rocks came up out of the tube like shot from a gun. The fellows learned to stay clear and warn others. The object from now on was to keep the bottom of the treme in the new concrete so water did not get back in till moving to a new location.

My last job on the diving scow, after the piers had reached their final depth, was to run a cable from the east pier to shore for telephone service. Later, when erecting the towers, they tried walkie-talkies, not at all satisfactory in foggy weather. Hand held transceivers have come a long way since then.

After the diving operation was finished I worked as a carpenter, stripping the forms from the tower pedestal. My crew for that job had a few characters one can never forget. We had two Whiteys, both outstanding workers. To distinguish between them one was called Dishface Whitey and the other Gashouse Whitey. I have no idea who named them but the names sure stuck.

Dishface was giving signals on the derrick doing the stripping. We had hooked on a fair sized load, which was to be lowered and turned loose in the water. There always were beachcombers to tow the scrap lumber to one of their projects, and I have no doubt some of it went into the nearby Salmon Beach homes.

One ambitious young rigger was anxious to ride the load down, unhook it and ride back on the headache ball. The operator could not see the load when it reached the water, so he had to take signals from Dishface Whitey. As soon as the rigger turned the load loose it drifted away, and he was sitting nicely on the headache ball hanging onto the lifting line. But instead of going back up, he was going down! He stood up on the ball and tried to climb the line. Whitey wet the young man's feet, then slowly lowered him till he was waist deep before signaling the operator to bring up the line. It was a warm day and all done in fun, but that water was cold.

Bridging The Narrows

My work for Pacific Bridge finished, I went to pile-driving again and eventually, in 10 years, back to the Narrows bridge to lay the new deck steel. A question asked of me many times always makes me think of the fine crews I worked with on both bridge jobs:

"Just what sort of a person becomes a hard hat diver, or a high structural steelworker?"

I will say he cannot be timid, must be in good physical condition. His fast reaction could save his life, or the life of someone else. He should always think of the safety of his partner as well as his own. A sense of teamwork and cooperation is a factor also. Beyond that, I would say it is just a fair cross section of our population that finds satisfaction in seeing some structure completed, that will stand for years. My crew worked on the diving scow for Pacific Bridge in all kinds of weather with never a grumble. They were Fred Mickelson, Marion Christian, Bill, Chet and Ernie LaChapelle, Dick Trussler, Ernie Warfield and Slim Haggerman.

And those piers we helped to build withstood the collapse of Galloping Gertie four months later, and the severest test of all, the 1949 earthquake, with no damage whatever.

Walt Cathey watches as a clamshell brings up earth and concrete rubble from the bottom of each cell, the final operation in sinking the piers to their prescribed depth—55 feet into the bottom for the west pier, 90 feet for the east pier.
The Peninsula Historical Society collections

The clamshell shovel cleans out the bottom of the pier, now set into the earth, before the final concrete seal.
Bashford photo, The Peninsula Historical Society collections

About to empty its load, a bucket of concrete was lifted off the scow and will be dumped into the treme funneling it into the pier cell forms. Bashford photo, The Peninsula Historical Society collections

Bridging The Narrows

A pedestal is poured to cap the top of the pier, July, 1939. Bashford photo, The Peninsula Historical Society collections

Timber fenders finish the overhanging pedestal, and will deflect marine debris and traffic.
Bashford photo, The Peninsula Historical Society collections

Bridging The Narrows

Bridging The Narrows

Galloping Gertie's deck steel was prefabricated in sections having eight-foot high sides, barged to the site. Their resistance to wind contributed to her fall.
The Peninsula Historical Society collections

A pre-fabricated section of deck steel is being prepared for lifting, to become the mid-span of Galloping Gertie. Above, lifting gear can be seen to the right of each catwalk obscuring the main suspension cables, in place.
Bashford/Thompson photo, The Peninsula Historical Society collections

Chapter Four

BETWEEN BRIDGES

Fate has a way of playing strange tricks on us. Here I had just finished the most interesting job of my life, helping construct the piers for the proud new Narrows Bridge. But my next job, a small one having nothing to do with the bridge, was connected to it by a string of events.

After the bridge opened to traffic July 1, 1940, the powder company located at Fredrickson came up with the idea of trucking powder across the bridge and loading it on a freight boat at the no longer used ferry landing at Pt. Fosdick. This is where the hidden joker came into play, speaking of fate. The idea was to remove the pilings in such a way that a boat could lay crosswise and load at high tide.

At the time, the state did not insist on the piles being pulled to remove them. They could be shot off with dynamite instead, and usually about 60 percent was used. I had considerable experience with this, so Industrial Engineers & Contractors sent me out to get it done. There was so much seaweed on the piling that it was quite difficult to get the charge where it was needed, which was at ground level. At low tide there was still 10 feet of water swirling around the dock supports. After the first shot the dogfish swarmed in, smelling the pogies that were killed and feeding at the surface. I'd had no idea there were so many of them. What I was seeing and hearing was beyond belief—slashing teeth everywhere, dogfish eating dogfish as well as everything else, their snapping jaws as loud as the old steel traps we used when I was a boy.

I finished that job. In the meantime the bridge earned its label "Galloping Gertie" by her reaction to seasonal winds, and fell down shortly after, on November 7, 1940. A piledriver had to come in immediately and redrive the ferry landing I worked so hard to destroy.

Before Gertie fell I saw a billboard advertising a bank near the Tacoma end of the bridge saying, "Safe as the Narrows Bridge." You can believe that sign was taken down very quickly after the disaster.

Bridging The Narrows

But those piers my crew and I helped build are still there where we put them, supporting the present Tacoma Narrows Bridge. The only change has been to the pedestals topping the piers at water level, which had to be enlarged to provide a wider base for the new and heavier towers to sit on. The old pedestals were removed, and new ones were formed at 60-foot centers and to a greater height above water than Gertie's had been. Because salt spray had started corrosion on the old towers, the new pedestal raises the base of the tower steel to 32.5 feet above high tide.

I have mentioned the April 13, 1949 earthquake. It was measured at 7.1 on the Richter scale, equally as strong as the October 17, 1989 California quake, if not more so. Although it shook off a cable saddle from the top of the east tower during construction of the second bridge, again no damage was done to the piers that had already suffered the stress of Gertie's collapse four months and a week after she opened to traffic in 1940.

In his final report on the Tacoma Narrows Bridge June, 1952, Charles E. Andrew, Principal Consulting Engineer, said this about the piers:

"Seventeen recorded earthquakes had occurred in this area since the original piers were constructed in 1939. Five of these quakes were of 5 or greater intensity on the Mercalli Scale, while two in 1946 were rated 7. In addition to these earthquakes, the destruction of the original bridge undoubtedly imposed lateral forces and pressures on the foundation material of a much greater intensity than will likely occur under the new structure. Very accurate surveys since the failure have indicated no movement whatsoever, lateral or vertical, occurred in the main piers.

"After due consideration of all the factors involved, the Board of Consulting Engineers decided that no modification of the existing channel piers was necessary below the top slab."

Andrew's report also tells about the riprap put in around the west pier during the new construction. I came on the job in 1950 to lay the deck steel from the west pier to midspan, after the towers and cables were up, so I wasn't there when it was done. But I was told that the bottom had become deeper, and tidal eddying was stronger around the west pier than the east. I remembered the eddies and whirlpools around that west pier, but it still was a surprise to me, in view of the constant trouble we'd had with the eight to 10 mph tide flow, pouring the east pier. Andrew says that soundings taken in 1940 and again in 1946 showed scouring had occurred at the west pier. The riprap was included in the new contract to prevent that in future.

Galloping Gertie was designed very skimpily, in my opinion. A minimum ratio for cables of 1/30th of the span, center to center, had been established for suspension bridges in the past. Gertie's cables were figured at only 1/72nd. Engineering authority David B. Steinman says that what proved critical in Gertie was the vertical slenderness of her span and the relative shallowness of her stiffening trusses. A generation earlier, authorities were recommending a minimum depth of 1/40th of the span. This depth ratio later was reduced to range from 1/90th to 1/50th for spans between 2,000 and 3,000 feet. The stiffening girders on Gertie were solid, eight-foot deep girders on each side for the full length of the center span, 2,800 feet, or a ratio of 1/350th of the span.

Bridging The Narrows

Original plans for Gertie called for a central span of 2,800 feet, and side spans of 1,100 feet each, with an open stiffening truss 25 feet in depth. It is my further opinion that had this plan been used, chances are she would not have fallen. The Public Works Administration had approved a grant of 45 percent of the cost, but required the state to employ a board of independent consultants to check the plans, according to Principle Consulting Engineer Charles E. Andrew. As a result, the firm of Moran & Proctor was employed for the substructure, and Leon Moisseiff for the superstructure. The board itself had made the change in plans that substituted the shallow eight-foot girder for the 25-foot open stiffening truss.

Professor F.B. Farquharson built an $11,600 model of Gertie and subjected it to rigorous tests, when all those engineers would have had to do was to research bridge history for the answer they did not get. In 1849 Charles Elliot spanned the Ohio River with a bridge 1,010 ft. long, destroyed six years later by wind. The *Wheeling Intelligencer* described exactly the undulations that also destroyed Gertie. Even before that, on Nov. 30, 1836, wind undulation collapsed a chain link suspension bridge at Brighton, England.

One hundred years before Gertie was built, John Scott Russel, vice president of the Royal Scottish Society of Arts, published a paper entitled, "On the Vibration of Suspension Bridges and Other Slender Structures; and the Means of Preventing Injury From This Cause." In it the author discussed the nature of vibrations that could destroy a bridge. These examples all can be found in Steinman's book, *Bridges*. In my opinion, not researching in available information, or not learning from research done, was a rather expensive mistake.

While all suspension bridges have design elements in common, such as cables and towers, each one built has its unique features. The cables used in the St. Johns Bridge in Portland, Oregon, are spirally wound, pre-stressed cables, usually used on spans of less than 1,500 ft. Gertie's 5,000-ft. span demanded a "straight lay" cable, formed in strands that are anchored separately but combined before leaving the anchorage to create the large, 17-1/2-inch diameter cables that support the bridge. The present Narrows Bridge main suspension cables are 20-1/4 inches in diameter.

Today's Narrows Bridge has suspender cables, also called "hanger ropes" (but not by the construction crews), hanging down from the main suspension cables to support the roadway. These slender cables hang in clusters of four every 30 feet for the entire 5000-foot suspension length. Each cable in the four ends in a zinc piece called a jewel, measuring five inches by eight inches, that was heat-cast along with a steel washer in the Roebling factory. These jewels, when the decking was completed, were located along the undersurface of the deck, acting as stoppers on the ends of the suspender cables, and they carry the entire weight of the bridge.

Zinc has a relatively low melting point of just under 800 degrees Fahrenheit. When the second bridge opened to traffic one function of bridge personnel was firetruck escort duty. Every truck carrying explosives or inflammables was followed by firefighting equipment over the bridge, in an effort to protect the cable jewels holding up the bridge. This was reported in the *Seattle Post Intelligencer,* Mar. 17, 1965. In late years I have never found anyone who was aware of this. It is my belief this practice was discontinued when the tolls were taken off early in 1965. If this is true it leads me to believe that the companies insuring the bridge may have insisted on this provision.

Bridging The Narrows

The world's third longest suspension span conducted traffic across the Tacoma Narrows for four months and six days. This view looks into the south Puget Sound with the mainland on the left, the Gig Harbor peninsula on the right and Fox Island on the horizon. Bashford photo, The Peninsula Historical Society collections

Bridging The Narrows

Towers are stacks of hollow cells. Individual sections of Gertie's towers were 32 feet long and weighed 24 tons, compared to 27 tons in the present bridge and 35 tons in the Golden Gate. In some cases towers are tall and flexible to allow movement to compensate for varying loads on different parts of the span. The two main suspension cables, running from end to end of the bridge, are supported across the tops of the towers. The present Narrows Bridge has rollers installed under the cable saddles which, as could be expected, reached the point where they refused to roll.

The present maintenance crew has developed a method by which they were able to drill and insert alemite fittings, so lubricant could be forced into the critical points. From past experience I know most rollers do not work long without an adequate grease supply. Some 66 years ago I helped install them on King County steel bridges, and found them frozen a few years later.

Expansion and contraction of its metal parts must be planned for in the design of any bridge. Most people do not visualize a metal tower moving under the full impact of a summer sun on one side. The expansion causes the tower to try to move away in the opposite direction. On the other hand, I recall an incident using the factor of contraction to good effect when crews were installing Gertie's last section of deck girders, even though it reflected a bit of a mistake in the sliderule mechanics of that day. I have never seen this following incident described in print:

I was on other work at the time, but an operator friend, Elmer "Red" Johnson, one of the best operators west of the Mississippi River, kept me well informed on the progress of the towers and deck steel. He said when the girders were to go in place it was found they were just a bit too long. They were allowed to cool all night, and the men tried again in the morning. Still they would not fit. A.S. Halteman, resident engineer and Grover McClain, erection superintendent for Bethlehem Steel, were discussing this when Andy Zori, an assistant superintendent, spoke up.

"You guys have tried your way," he said. "If you're willing to turn it over to me I'll put it in." They were willing, so Andy ordered plenty of dry ice to pack around the two beams, which again sat all night. The next morning Andy slipped them into place as easy as putting butter on bread, reported Johnson. It was the perfect solution. Everyone was happy and did not mind paying the sizeable bill for dry ice. I am sorry to add that Andy Zori lost his life shortly after on a railroad crossing near Tenino, where an automatic signal failed to work.

Just prior to the fall of Gertie, Hart Construction's steam piledriver was working on the storm cables being rigged under the east end of the bridge when a painter working on the bridge knocked a paint can overboard. With all that space to fall in, it hit the head of a longtime friend of mine, Hughie Mickeljohn. His death was instantaneous.

Not long after this, on the day that Gertie fell, the same crew and piledriver were in the water under the west end of the bridge, between the pier and the shore. They were attempting to stabilize the violent undulations going on in the deck level by driving anchors for storm lines. Bud Brown told me that chunks of concrete were popping off and falling all around them, and they wondered if they would live through that ordeal. Fortunately they were some distance away from the collapsing section when it let go. Besides Bud there were foreman

Bridging The Narrows

Gertie writhed more than usual, on Nov. 7, 1940. At 10 a.m. winds had reached 42 mph and by 10:30 collections of tolls was discontinued. This photograph made a popular postcard. Howard Clifford photo. The Peninsula Historical Society collections

James Bashford's well-known photo shows the rigid undergirding that contributed to the bridge's fall at 11:08 a.m. A Hart Construction Co. piledriver can be seen in the water near the west pier, to left. Attempting to put on storm cables, it was slow getting out from under. Bud Brown, a worker aboard, described chunks of concrete falling near.
Bashford photo, courtesy The Span Deli

Bridging The Narrows

Lyle Lewis, operator Charley Burgeson and Julius Nelson, whom I recall living through this episode.

Because I knew how well and carefully built Gertie's piers were, I was not surprised to hear later that they were unharmed in the crash. There was some talk by authorities of simply cutting the suspension strands at the anchorages to let Gertie's remains fall into the Narrows. However, the war in Europe was creating a need for scrap metal, and within a month after the collapse the J.H. Pomeroy Company was contracted to remove the balance of the broken deck to clear the way for salvaging.

In August, 1942, my friend Ralph Riffe joined the crews hired by the J.P. Murphy and Woodworth Co. to pull the wire cables and take down the towers. This work was considered extremely hazardous besides difficult, but was accomplished without any loss of life. Ralph was an excellent operator, one of those easygoing guys whom nothing ever bothers. He and I had dredged together all one winter with a floating rig for Simpson Timber Co., and had worked together on other jobs for many years.

In the final view of the failure of Gertie's design, its self destruction was a blessing. We could have been stuck with a totally inadequate bridge of two lanes. The present bridge with four lanes is inadequate now, at peak traffic hours. Commuters often experience a slowdown to a crawl, sometimes a complete stop, and perhaps with all four lanes filled with cars for its total length of 5,979 feet. In those moments, it helps to have read Andrew's report, which says the main cables are capable of supporting a minimum of 220,000 pounds per square inch.

Salvage was directed by cost-plus contracts with J.H. Pomeroy & Co., December 1940, and with J. Philip Murphy & Woodworth Co., August, 1942. They retrieved the steel in towers and deck, and wire from the frayed cables. Bashford photo, The Peninsula Historical Society collections

Bridging The Narrows

Ralph Riffe, the author's longtime friend and co-worker, worked with crews salvaging materials from Gertie during WWII. A heavy equipment operator with an easygoing temperament, Riffe could make repairs with scrap materials that worked as well as a new part.
Bashford photo, The Peninsula Historical Society collections

Chapter Five

TOWERS

The construction of most modern suspension bridges will take place in a prescribed procedure. After site preparation, appropriate crews adapt their methods to the local environment and undertake their jobs in a sequence of separate completions, described here:

1. Over the location of a pier, blocks of reinforced concrete called anchors are dumped into the water as onto a target, then secured with cables to the pier as it is being constructed.

2. Sinking the piers is a gradual process, building with steel and concrete onto the initial pre-formed, hollow caisson. As it "grows" upward and sinks downward it is stabilized in place by cables to the surrounding anchors. The caisson's steel cutting edge will sink into the earth. The pier is then excavated to the proper depth through its hollow cells. Heavy concrete seals bottom and top, before a broad concrete pedestal caps the top. An added timber fender deflects marine debris and traffic.

3. A tower is erected on each pier pedestal, and finished with a steel saddle on its highest point to receive the main suspension cable.

4. The two main suspension cables are spun from wire spools unreeling onsite and affixed to an anchorage built on land at each end to receive them. Temporary catwalks for workers are placed from tower to tower so the cable may be spun and laid in the air across the towers to the second anchorage.

5. Individual suspender cables, also called hanger ropes by their manufacturer, are now hung at prescribed intervals along the full length of the two main cables. These will hang in air until attached by zinc "jewels" to each section of the deck as it is completed.

6. Building the deck begins with placing the pre-formed steel sections, piece by piece. Four crews begin back to back at each tower, two working toward each end and two toward midspan. When all steel sections are riveted into place the deck is ready to receive a surface.

7. Dampeners are the final element of steelwork in the construction of a suspension bridge. These are hydraulic units attached at points on the underside of the roadbed. They act as shock absorbers to slow down motion in the structure caused by traffic, wind or earthquake.

8. A surface is applied to the roadbed, which in the case of the TNB included four traffic lanes, three 33-inch wide metal gratings the length of the span, and a 19-inch grating against each of two sidewalks.

Before the present roadway design was accepted, a test slab was constructed exactly as the bridge floor would be. Strain gauges showed the slab, with an approximately 135 percent overload, to have no flaws whatever. The test concrete showed a compressive strength of 4,500 pounds per square inch after 28 days. The concrete of the finished roadway weighs 118 pounds per square foot, with an asphalt wearing surface of 5/8ths inch.

While the piers were being sunk for Gertie, work went on around the clock. Construction of the second Tacoma Narrows Bridge, built upon the original piers, was done in regular eight-hour daylight shifts. I came on the TNB job after the work described above in steps 1 through 6 was completed. I am therefore indebted for information to Al Sonn, Puyallup, who not only worked on Gertie at the time I did, he also was a foreman on the towers and cables for the present bridge.

Each tower presents two legs, one each side of the roadway. The structural components of each section of leg are steel cells each 32 feet long, stacked upright like crackerboxes in groups of four by the creeper derrick and connected, end to end, to the section below it with one-inch rivets. The creeper derrick (also called a jumping derrick) was capable of lifting 80 tons while attached to girders securely fastened to the tower. The group of four cells in each leg were riveted together in a shape resembling the Red Cross symbol, and each of these groups made a "lift," the term for the maximum height reached before the creeper derrick moved up the tower another 32 feet. The four hollow cells are bound together internally by steel diaphragm plates every seven vertical feet. Every 32-foot leg section of the tower thus formed weighs 27 tons. Upon being connected by X-bracing the two legs become a structurally sound unit: a tower, with an elevator running the height of one of its leg cells, and a roadway running through its lower elevation.

Outriggers, as we called them, buttressed the towers to stabilize them while under construction. These temporary devices were removed from the towers after the main suspension cables were completed and in place.

The steel cable saddles weighed 28 tons, not 23, as reported in the *Tacoma News Tribune* April 14, 1949. Each saddle is fitted in the top of the tower to receive the main cable laid across it, and is held in place by 36 one-inch bolts. On the TNB, rollers between tower and saddle allow stress movement.

The suspender, or hanger cables came from John A. Roebling and Sons of California, finished to specified length with the zinc jewels cast in place at the lower ends. They are hung from the main suspension cables before the deck steel is laid, waiting for attachment through and under the deck.

Each main suspension cable was formed on the job, spun from 19 strands, each strand composed of 458 No. 6-gauge wires. Where the strands pass over the curved saddles at the tower top, zinc fillers approximately six inches long and strung on wire rope fill the interstices between the strands. This cushion is meant to keep the strands from deforming. The two main suspension cables provide the ultimate stabilizing force for the entire bridge.

The 19 endless strands of each cable are splayed at either end where they attach to their east and west anchorages. Each of the strands passes around the "shoe" end of an anchor bar that is embedded 62 feet into the concrete anchorage. This adjustable shoe is similar to the rigger's shiv, or pulley, and provides a minimum inside diameter of 26 inches for the strand's loop around it.

A steel baseplate for the new tower is placed on a rebuilt pedestal designed on a 60-foot center. Distribution of load to main walls of the original piers is improved. From pamphlet: "Souvenir of Tacoma Narrows Bridge", Pioneer, Inc.

Bridging The Narrows

The first leg cell is lifted to be placed on the west pier pedestal. From Pioneer, Inc. pamphlet

Jack Hamilton, a superintendent with Bethlehem Pacific Coast Steel Corporation, directs the lift of a tower leg cell. The rigging here shows a good example of how two sheaves (top), cables and four shackles equalize strain and weight.
From Pioneer, Inc. pamphlet

Bridging The Narrows

Setting it down, pushers Red Turner, bending, and Al Sonn, right, work with a Washington state inspector on left.
From Pioneer, Inc. pamphlet

As the tower rises, temporary steel outriggers buttress it against wind motion. Jack Durkee photo

Hollow X-braces were pre-formed and assembled onsite by workers, at times suspended in midair, with the aid of the Chicago boom. Jack Durkee photos

Bridging The Narrows

This saddle, designed to hold the main suspension cable where it passes across the tower, tumbled into the water during the April 13, 1949 earthquake. Here it is being recovered April 15 after being located by a diver. Note the plume of smoke from a Northern Pacific steam engine, east bank of the Narrows, in background.
Jack Durkee photo

Workers have placed the saddle to receive the main suspension cable atop the west tower, and are fitting the guide wire for the main cable. From Pioneer, Inc. pamphlet

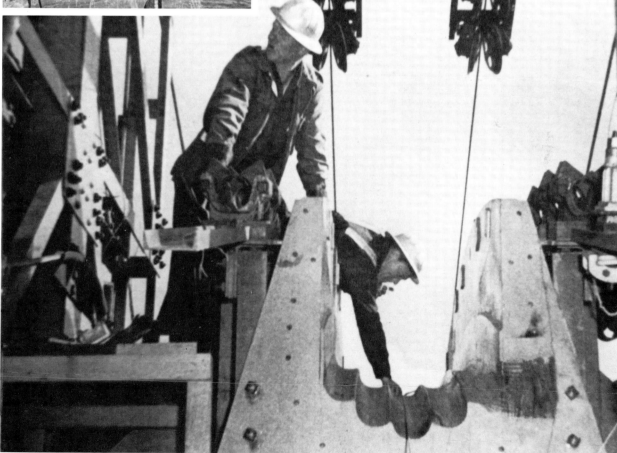

A catwalk spans the space between towers for workers who will spin and install the main cables. From Pioneer, Inc. pamphlet

Cable workers Warren D. Medak and John Wall put out their gear to prepare for spinning the main cable.
From Pioneer, Inc. pamphlet

Bridging The Narrows

The gear for spinning the cable is in place and creation of the continuous, multi-strand cable can begin.
From Pioneer, Inc. pamphlet

From the catwalk, Al Sonn measures the completed cable, 20¼ inches in diameter. It provides the towers' primary stabilizing influence. Joe Gotchy photo

Bridging The Narrows

These eyebars, holding strands of the suspension cable looped around 26-inch "shoes," were part of the anchorage for Galloping Gertie. A similar assembly replaced them for the new bridge, extending 62 feet into the rebuilt concrete. Bashford photo, The Peninsula Historical Society collections

A new west anchorage replaces the old one, utilizing part of the original shell. The new concrete was poured in alternating blocks of 150 to 220 cubic yards. Bashford photo, Pioneer, Inc. pamphlet

Bridging The Narrows

The crew of the west midspan traveler posed on April 26, 1950. From left, Doyle Dunn, Joe Gotchy, Hal Mousseau, Bill Vogle, Hank Meir, H. Woody Wood, Hap Shaffer, Frank Butler. Jack Durkee photo

Chapter Six

DECK STEEL: 10 YEARS LATER

In 1950 I was working on the Capitol Lake dam in Olympia until the towers and cables of the present Narrows Bridge were finished. Then I reported in at the union hall where the dispatcher said, "Joe, we have just the job for you. I hope you don't mind the risk of being canned. They've canned two men on the rig I'm sending you on."

The first morning on the job, after a boat ride to the west pier, an elevator took us to deck level. When we stepped out on a platform I saw we had to walk the steel, as no wooden walkway had been installed yet. This gave us two choices: We could walk on the top chord at the outer edge and dodge the suspender cables or walk the deck stringers, which were I-beams eight inches wide and about 18 inches deep. Some men straddled the stringer, walking on the bottom flange, which seemed prudent. But I saw a few walking along the top of the eight-inch beam and decided to test myself. It had been eight years since I had so much space under me, but after the first step no problem existed for me. I even felt some satisfaction in knowing I could still feel as safe as ever.

When I arrived at the west midspan traveling derrick the superintendent, Jesse Cheeley, said to me, "I hope you don't mind if I sit beside you for a bit."

"Make yourself at home," I told him. In about 20 minutes he stood up and said, "I'm leaving, Joe. You are O.K."

At the time, I did not know this crew had come intact from Los Angeles and were without a doubt the best to be had, with plenty of experience in bridge work. Frank Butler, my foreman, and the rest of the gang accepted me as one of them. Hap Shaffer, signalman, made my work easy. These rigs—travelers, we called them—had no locking foot brakes, which meant that the operator sometimes had to hold a load by foot pressure for 20 minutes or more. This could be quite tiring, but Hap, whenever possible, would give me a signal to dog the load, and that meant I could hook the dog into the notched drum and rest my leg.

Bridging The Narrows

The job proved to be one I could carry a camera on with me, and get some interesting pictures. Frank liked my pictures and used to say, when my traveler was being winched to a new setup, "Joe, you have 30 minutes to go and take pictures." I did take full advantage of this, but they never had to wait for me. I also carried a small, six-power scope with a clip so it could be carried in a shirt pocket. It wasn't long before Frank Butler discovered my "peeper," as he called it. About two or three times a day he would borrow it to see what our competition, the east midspan gang, was doing.

The deck steel for constructing the second Narrows Bridge was fabricated at the Bethlehem Company's plant at Pottstown, Pennsylvania. Wooden templates, similar to those used in ship construction and made to the bridge's specifications, were laid out on the steel so that every rivet hole matched. The deck panels were assembled prior to leaving the plant; on passing final inspection they were disassembled and shipped to Tacoma.

The steel was loaded on Foss scows at the Port of Tacoma and towed the nine nautical miles out of Commencement Bay around Point Defiance and south to the bridge site in the Tacoma Narrows. Chicago booms mounted on the sides of the two towers, well above deck level, picked the steel and materials off the scow—first to construct the first section of deck, next for the on-deck assembly of the four traveling derricks, then, every piece of steel needed by the traveler crew for span construction. From each tower one traveling derrick would work toward shore and one toward center span, as the deck steel was laid in place.

The travelers were built of a heavily reinforced steel frame approximately 30 feet wide and 40 feet long. Like the little material cars that supplied them, they had flanged wheels that rolled on the deck stringers as if on rails. At the back of each were two jacks that allowed the traveler to be adjusted, to be always level as it slanted upward toward the center of the span or downward toward shore. At each new setup for another section, steel clamps held the traveler solidly to the deck stringers. When the traveler was moved to a new setup, all I had to do was boom up and secure all lines; a line from the material car moved the traveler. Ted Joslyn was flagman and rode the car, moved by its own gas engine-powered winch.

My traveler was a three-drum rig with swing gear in front. It housed a large Climax engine and a winch that consisted of a boomdrum, a heavy load drum and a "whip drum," also called a light load drum. There were two swing drums, as well. This traveling derrick was called a "layleg" rig. Two laylegs were pinned to a mast, allowing the boom to swing more than 180 degrees. At the bottom of the mast was the swing gear, a steel circle to which a line was fastened from the two swing drums. About 12 tons was the heaviest lift, made when the derrick had to lift up each side of a completed new setup of deck steel. When the section, about 60 feet long and the width of the bridge, was finished, and before the traveler moved ahead, the deck was lifted up twice, a few inches on each side, in order to hook up the jewels and cables that would suspend it.

Inside at the controls, behind a canvas windbreak, I worked the traveler placing steel with only the lights on the board in front of me for guide. I never saw the work as it was happening, until warm weather in May, and could not see hand signals, only the lights my signalman sent my board. But with a good signalman, which Hap Shaffer was, this system was more than adequate.

Bridging The Narrows

A Chicago boom, attached at an angle to the tower leg, lifts supplies from the material scow below. Open deckwork appears on right. The west midspan traveler, center, is beginning its first setup. Jack Durkee photo

Inside the traveler derrick Joe Gotchy dogged the drum for Jack Durkee's camera. Jack Durkee photo

A floor beam is transported to the derrick on the materials car. Jack Durkee photo

In place, floor beam fills the width of the bridge. Suspender cables with jewels attached hang from main cable, awaiting connection with roadbed. Jack Durkee photo

Hal Mousseau drives a steel and manganese pin to align rivet holes of a floor beam with those in the top chord. Joe Gotchy photo

Woody Wood, left, and Hank Meir, right, replace pins with bolts from basket carried to their point. Riveters will follow. Joe Gotchy photo

Bridging The Narrows

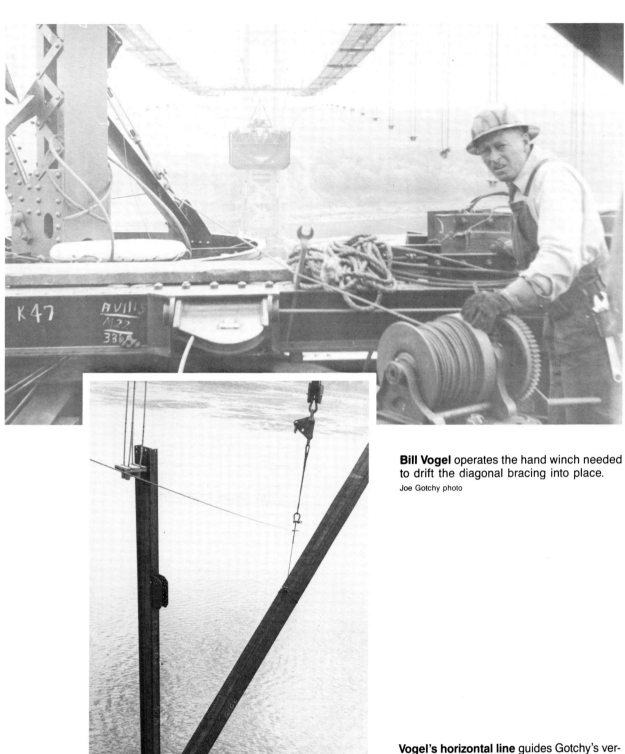

Bill Vogel operates the hand winch needed to drift the diagonal bracing into place.
Joe Gotchy photo

Vogel's horizontal line guides Gotchy's vertical line, holding a diagonal truss in place while Hal Mousseau and Hank Meir connect. Both beams will support another top chord, into which the suspender cables, seen here hanging free, will be inserted. Note jewels terminating the four cables, upon which the weight of the deck will rest. Jack Durkee photo

Bridging The Narrows

I've operated many cranes and derricks, a few with fluid drive similar to the device that moves the transmission in a modern car. But for constructing a suspension bridge such as the Narrows I would still prefer the old three-drum side lever rig with swing drums in front. Regardless if one is lifting, lowering or swinging a load, smoothness is a neccesity for providing a crew of riggers with a feeling of safety. If the operator is not moving his rig smoothly he won't be with that gang long.

To be sure of smooth handling on the light load drum, or whip, as this single line was called, I had to start up my rig five minutes early in the morning. I ran the line up while riding the brake, to heat the brake band and dry out the lining. The headache ball (slack puller) would bring it back down. A few trips like that dried out the dampness that always accumulated from the night air; otherwise the braking action was very undependable.

To lower a load of a few tons on the heavy load drum with a steady momentum, I used to dog the boom drum and put pressure on its brake so the boom drum was securely locked, then throw out the engine clutch while holding in the lever on the load so the gears could run backwards, acting as a governor. Holding the brake on the lowering drum let the load down with very smooth action. I could also use the lever on the boom drum as a brake as long as the dog was in. I always used this method when picking up a completed deck section to hook up the suspender cables.

The first section of deck in each of the four work stations was built with the aid of the Chicago boom. Then, when there was a footing for the traveler to work from, construction of the deck went ahead in roughly four steps.

First, the outer edges of the bridge span were formed by placing the top and bottom chords to create a depth of approximately 33 feet and a length of 60 feet to be filled. The top and bottom chords broke joints with each other, similar to the way bricklayers line up two courses of bricks. Diagonals and braces connected across the chords. Next, the pre-trussed floor beams are lifted up and set at right angles across the space between the two sets of chords. They also will have steel bracing added in and connected. Third, the deck stringers are laid down lengthwise with the bridge. These, as I have mentioned, are steel I-beams eight inches wide and and 18 inches deep, and add up to eight across the width of the roadway plus one under each sidewalk.

While the deck stringers are the last steel to be placed, the next-to-final step in completing a section of deck is the slight lifting of the entire structure, about a 12-ton lift. As I picked up each edge a few inches, the men placed and secured the zinc jewels hanging on the ends of each group of four suspender cables, thereby providing the permanent means of deck support. Then we laid the steel deck stringers, and moved ahead to the next setup.

Shortly after the bridge photo, next page, was taken, disaster struck. A young man, Stuart Gale, who had worked on the towers and cable, went to work with the west shore span gang. A certain brace he had noticed, that depended on a weld to hold it till it could be riveted, made Stuart uneasy. He wanted bolts in it. On his third day on the job, he and steel fell 180 feet into the Narrows. I was on the west midspan traveler, shown on right in the photo. We got the news instantly: A warning whistle blew. I looked down and saw Stuart swept past the anchored material scow by the fast-flowing incoming tide, and sinking rapidly. A husky young ironworker, Harold Peterson, unhesitatingly dove in, but could not reach him. The belt of tools on Stuart made the difference, I'm sure. By the time the patrol boat reached Peterson he must have been a quarter mile away. With whirlpools at the pier, it would take a lot of courage for a strong swimmer to duplicate this. As soon as Peterson was brought back the job was closed down for the day—a tradition I never knew to be broken in those days. Needless to say, bolts were used after this fatality. The safety meeting, later, came up with a possible cause of the faulty weld: Maybe someone stopped for lunch, or went to use the latrine, and forgot to finish it.

I saw one more bridge worker fall into the Narrows. With those belts of tools they went down so quickly nothing could be done. The last man was lost when the two midspan travelers were close enough so that we could easily watch and recognize each other. I was holding a load being connected, and as I glanced at the other gang I saw everyone looking down at the water. I looked, too, and saw a man sinking; I did not see him move. His clothing gradually took on a greenish hue as he went deeper and finally disappeared. He was from the east midspan gang—Whitey Davis, if I recall correctly. A few days later Bethlehem stopped work long enough to perform a funeral service on board a boat carrying his wife; she cast a wreath on the water below the bridge.

The west midspan, right, takes shape at the west tower. Stuart Gale's fatal fall was from the deck portion extending toward shore, left. The derricks of two travelers can be seen, left and right, and of the Chicago boom attached to the tower. Temporary outriggers, based on the pedestal, have not yet been removed. Jack Durkee photo

Bridging The Narrows

Doyle Dunn waits for the diagonal he will connect to the top chord, upper left. A group of four suspender cables has been secured by their jewels through an opening in the chord. Jack Durkee photo

Bridging The Narrows

Hank Meir, on the bottom chord, connects more brace pieces. He will pin and bolt; the riveters will follow. Joe Gotchy photo

With the towers up and the cables in place, the traveler crews lay deck steeel. Here the west midspan traveler appears in foreground, west shore traveler rear. The traveler derrick has just picked off a deck stringer brought forward on the materials car, left. All three rigs roll on flanged wheels on the deck stringers, seen between them. Jack Durkee photo

Gotchy's line lowers a deck stringer, the last piece of steel in a deck section. View is to the east shore. Jack Durkee photo

Ted Joslyn rides the chord on his materials car signaling to the operator of the "railroad." The steel chord will shape the outer edge of the bridge. Jack Durkee photo

Bridging The Narrows

Bottom chord for midspan is picked off the car by the traveler and swing into place for connecting. _{Jack Durkee photo}

Bridging The Narrows

Foreman Frank Butler stands in front of west mid-span traveler with Al Sonn, foreman on towers and cable, and Hap Schaffer, signalman. Joe Gotchy photo

Work progresses on both sides of the west tower. A scow loaded with a trussed floor beam, visible on left between catwalk and deck, approaches the pier. Jack Durkee photo

Bridging The Narrows

Down for lunch in the lunchroom scow at the west pier; a Bethlehem Steel Co. patrol boat is on duty. Point Fosdick, background. Jack Durkee photo

Bridging The Narrows

Casualties on the second bridge numbered three, I've read, and only recently did I hear of the death of Robert E. Drake, who was killed May 24, 1948. His son, Wes Drake and step-son, Frank Nimick, both of Tacoma, told me about this. Drake was working for Woodworth Co., who had a subcontract for strengthening the anchorages, placing riprap for the west pier and enlarging the pier pedestals when the first work on the second bridge began in the spring of 1948. He was on the ground below a derrick when a crane cable snapped, dropping the boom directly onto him.

One warm afternoon when we were hanging steel I had an experience I never want to repeat. I was holding a heavy load on the foot brake, regular practice when connecting. Usually, if the drum was dogged, the load was either too low or too high to line up the holes, so most of the work was done using the foot brake. I was thirsty, so I picked up a thermos of fruit juice sitting near me, and as I tilted my head back to swallow I became choked. Could not breathe. Knew something had to be done quickly or steel and ironworkers would drop into the Narrows 200 feet below.

I concentrated on easing the load onto the dogged drum, and stood up. On the point of passing out a thought struck me, "There must be some air in there yet." I folded my arms tightly across my diaphragm and threw my chest toward my knees as hard as possible, which popped my epiglottis open and I could breathe again. This happened years before Dr. Heimlich and his maneuver was ever heard of. It sure saved me, and I have no doubt had been done by others before me, though the thought came to me only as a last resort. I never did tell the crew of this, as I did not want them to worry about their safety.

The west midspan traveler crew consisted of eight men, including myself: Frank Butler, pusher (foreman); Hap Shaffer, my excellent signalman; H. "Woody" Wood, Hank Meir, Bill Vogel and Hal Mousseau, a big, goodlooking Frenchman about 6'4". Hal was the only man who never paid me for the pictures I let him have. He seemed to think that film, photo paper and enlargers were there strictly for his benefit, and that I should be happy to give them to such a nice-looking Frenchman. And there was Doyle Dunn, a good worker and a very quiet person. I heard he was killed on his very next job. These men made an excellent crew. Several times we were shut down for as much as 1-1/2 hours when we got too far ahead. On two weekends I had to go over to work on the east midspan traveler to help them get caught up. After that session, I understood why we were always ahead, and what a difference a good gang made.

The riveting crew worked behind my deck steel gang. Riveting the bridge together was a complex process requiring four men to place each rivet. And even before the riveters came on the "point," the place for the day's job, at least two men had worked the same spot before them.

After my rig placed a piece of deck steel, the first step in connecting it was to insert a steel pin through the rivet holes to make sure the alignment was exact. Next, a bolt was installed, firmly, but only as a temporary hold, in a minimum of 50 percent of the rivet holes. The workers who did this preparation for the riveters often straddled the steel and worked over empty air.

Then, wooden staging had to be set up to provide a safe working platform for the rivet

Bridging The Narrows

crew. Two or more "needle beams" (usually a good grade of 4x4) were hung from manilla rope in such a way that planking could be laid on them for the next man, who removed the bolts, and for three of the riveting gang—called the catcher, the riveter and the bucker-up. The first man on the rivet team was called the heater.

This man was usually located where he could throw a heated rivet from his tongs, not more than 30 feet, to the catcher (also called a passer). The heater had to know the number, size and length of rivets to be used at that particular point. He arranged his rivets on his forge— usually gas-fired, but they used to use coke—so they would be at the right temperature when needed.

(On my early bridge work some of our rivets were too cold and, if driven, never were tight enough. An inspector could tell by tapping them with a small hammer. If they didn't sound right we had to cut them out. If we rolled the gun on them, called "caulking," they might look good but were not a tight rivet.)

You might say the heater had the lead role in a team performance. His aim with hot rivets was crucial to the team's success. Sometimes, if the point was too far from the forge, he dropped the rivet into a pneumatic tube and pressed a lever conveying it to the driving gang. The catcher, who sometimes used a hand-held funnel to receive the rivet, and the riveter usually traded off so that each took a half-shift on the riveting gun. The bucker-up put himself behind the steel being riveted, often with his head into tight, difficult angles, and did his work with a hand tool or an airjack, to force the hot metal piece into a sure fit.

To put it more simply, the heater prepared and threw the rivets; the catcher received them and put them into the holes; and the bucker-up was ready with his tool on his side of the cold iron so the riveter could start driving the hot rivet immediately.

A riveting gang shows George Hicklin, left and L. Cox, center. Man on right wears the hat of a bucker-up.
Joe Gotchy Photo

Jack Durkee Photo

Riveters worked behind the deck-steel crews. Here, Leonard Dean, the heater, keeps both hands busy: his left tends to the forge while his right sends a hot rivet through the tube to the crew waiting on the point. Joe Gotchy photo

Dean, heater, aims a rivet from the forge to the catcher, who is ready with a catching cone. Joe Gotchy photo

George Hicklen drives one with a rivet gun. His bucker-up is below with head and tool inside the steel chord. Joe Gotchy photo

Bridging The Narrows

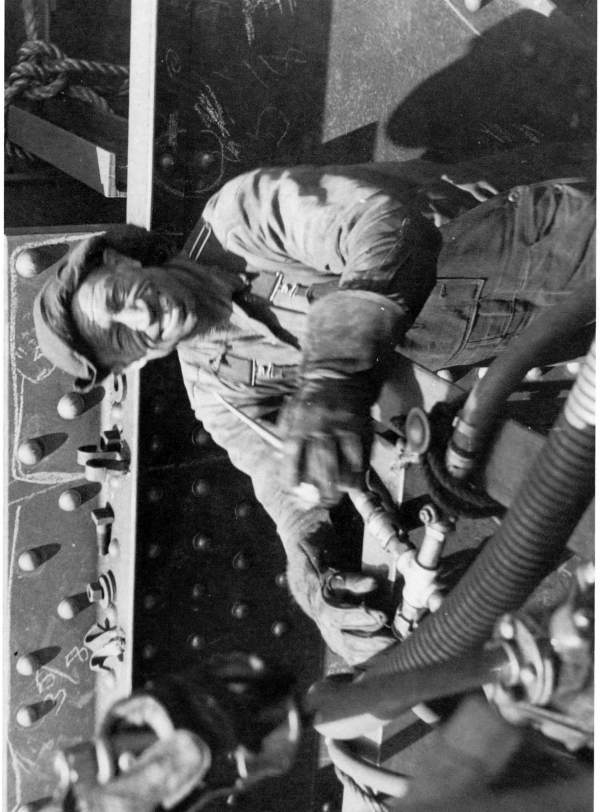

Pinetree Colby, riveter's bucker-up extraordinary, comes up for air. His work usually was performed inside tight places behind the riveter, necessary to form a head on the hot rivet before it cooled. His crumpled soft hat cushions his head against the steel and is the badge of his job. Joe Gotchy photo.

Bridging The Narrows

In a group of men like the ones who built the Tacoma Narrows Bridge there will always be a few who are unforgettable. I would like to introduce you to the one and only Pinetree Colby. When I first saw him he was so homely that I felt compelled to take his picture. But after seeing him at work in many unfavorable conditions, always with a smile on his face, I realized, "Here is a diamond in the rough." Believe me, it is not exactly a pleasant experience to put your head and shoulders into some of the places rivets must be placed, and endure the noise and dirt jarred loose by that chattering rivet gun, and to hold the rivet in place till a head forms on it. Pinetree always came up smiling.

I was very surprised, when I showed my picture of him to some of the gang and the word spread till I was swamped with orders for his picture. I had been off the job about two weeks when Bill Grant, one of the state inspectors, came to ask my permission to have Pinetree's picture printed in the *Tacoma News Tribune*. Pinetree had been raised on a Wyoming ranch and a step-sister, who had lost contact with him, saw his picture in the paper with my name under it. She called my wife, who told her how to contact him, and a happy reunion followed. She called again to thank us for making the reunion possible.

Scully Webster was another man not easy to forget. He was not only an operator furnishing the motive power for the material car at the west pier, but was also an entrepreneur. One day my rig was shut down waiting for more steel to be hung on the east span to bring us back in balance. I was taking pictures as I walked back to the tower. Was I surprised to see that Scully had a regular little canteen at his scene of operation. He had quite a display of cigarettes, candy bars and gloves for sale to any bridgeman in need of them, really a needed service. Evidently the Bethlehem Company thought so too. They did not put a stop to his selling, for it did not interfere with his work. But by the time I discovered Scully's canteen that day I had run out of film.

With so many operators on the job we often held a pseudo union meeting on the dock some mornings discussing the job. Paul Petrie, Jim Bigelow, Scully Webster, Pewee Johnson, Johnny Green, Louis Bolton and I were not too happy that we received 25 cents per hour less than the steel workers whose lives we literally held in our hands every day. Our wage scale, applied to all jobs started prior to 1950, was $2.33 per hour. That $1 is worth 17 cents of our present inflated currency. We agreed one morning we should discuss this with Tommy Martinsen, the resident engineer for Bethlehem Steel on the bridge job.

As it happened, I was the one who brought this to his attention. On the way down in the elevator as we were going to lunch I mentioned this wage matter to him. He agreed that we were underpaid, but he also said, "Joe, blame your union. You could have had more money, but this is the kind of agreement they signed and I can do nothing about it."

So many incidents happened in the construction of the second Narrows Bridge that could be woven into a very deserved eulogy to Thomas Martinsen. He had the respect of all the men I was acquainted with. Tommy did everything he could to keep the job moving, even in an earthquake. When a cable saddle was flipped off the east tower, before it was bolted into place, by the April 13, 1949 earthquake, it created a problem calling for immediate action. The 28-ton steel saddle fell into the water, hitting the rigging scow as it went, taking three compressors and scads of equipment with it. No crew was lost, and they'd had enough presence of mind to put buoys on the compressors, which were recovered the next day.

Bridging The Narrows

Scully Webster ran the tower winch for the materials car. He also supplied a successful "canteen" selling needed goods to workers. J. Bashford photo, The Peninsula Historical Society collections

Well toward center midspan, the main cable is too low for the derrick to work below it. The load line had to pass through the catwalk to attach suspender cables to a completed deck section. Joe Gotchy photo

This closeup of zinc cylinders, called "jewels," shows their placement snugged up against the steel flanges of a vertical post in the under-deck trussing. A steel washer comes between the flange and the zinc, which terminates each suspender cable. Zinc is chosen for this load bearing function because of its durability and resistance to corrosion. Jack Durkee photo

Taken from inside traveler, this view shows four crewmen rigging the lines passing through the catwalk to lift top and bottom chords with their bracing to connect the jeweled ends of the suspender cables. Joe Gotchy photo

Men from the four deck crews pose near the end of their jobs in July, 1950. From left, Danny Lowe, pusher for west tower crew; Frank Trinko, boss of a rivet gang; Frank Butler, foreman, west midspan; Jesse Cheeley, superintendent of both west tower and west midspan jobs; Jack Hamilton, superintendent of both east jobs; Buster Thompson. Joe Gotchy photo

Bridging The Narrows

Tommy called the Walter McCray Diving Co. of Seattle immediately, to retrieve the valuable saddle, and then was able to get tools enough from other contractors to keep the job running. Art McCray, Walter's son, had inherited the business after his father's death, and he brought down his own powered diving scow. Art searched underwater for three days, and said he found a large hole before he found the saddle. A line from the tower lifted it back on the job. I had seen his father dive when I was very young, and this incident always recalls to me the tragedy that occurred on Walter's last job when a bridge pier collapsed in the Aberdeen area. Art had to bring up his dad's body.

Then there was the fire on the west pier on June 9 that same year. The Bethlehem Company's boat operator, Charles Brooke, was working on his boat at the 6th Ave. service dock when the alarm sounded about 11:40 p.m. He and Tommy Martinsen jumped in the boat and sped over to the west pier, where the creosoted timber fenders had ignited. They worked like slaves rolling full drums of gasoline off the pier and into the water, till the fire made so much headway they were forced back into their boat. According to the *Tacoma News Tribune* story, the Tacoma fireboat and Foss tugboats equipped with firefighting monitors arrived, but could do little to reach the tower top, as by this time the fenders were fully involved. The creeper derrick was almost a total loss, as were three winches and the Chicago boom that unloaded material scows. Fortunately no steel in the towers was seriously damaged.

The fire, estimated in a *TNT* story to have caused a $500,000 loss, was believed to be started by a hot rivet. It does seem that when hot rivets are used it would be appropriate to have a watchman on the job with some firefighting experience and proper equipment to control a fire that is just getting a start. I recently talked to Al Sonn, Puyallup, who was a foreman on the bridge. He said the damage by fire was not nearly as serious as was first reported, and that the condition of the derrick was not serious. Most of the damage was in the creosoted wooden fenders surrounding the pier, and they had to be replaced.

Martinsen had looked to me like a college kid when he visited the diving scow I was on while sinking the piers for Galloping Gertie. He was with us for one dive. I had forgotten I'd known him before, but when I wrote to him in 1984 Mrs. Martinsen answered that he had died just before my letter got there. She told me his first job had been on the Golden Gate Bridge, then to the Bonneville Locks. He was on several other jobs before he came to work on Gertie. This makes him a veteran of both bridges. He put his heart into his job, and sometimes had to pull a rabbit out of a silk hat to keep it moving.

A boat was always waiting at 7:15 a.m. at the Sixth Ave. service dock to take us to the Narrows piers. I was always early, with plenty of time to visit should a seat partner be so inclined. After a few mornings, I found he usually was a young field engineer for Bethlehem Steel, 20 years my junior, but from the first we seemed to click. Jack Durkee was an interesting young man, with a broad outlook on life. In the short time it took to assemble the deck steel, from March to July, we became well acquainted. I believe Jack wrote most of the reports in his position as field engineer and he also took many pictures with a Speed Graphic showing work progress.

When my own work was finished in July, 1950, Jack said to me, "I know you have taken many pictures, Joe, but if you wish I will leave my negatives with you. Print what you like and return them to me." Needless to say, Jack had taken many pictures that I could not, and I was pleased to accept his offer. But I was very sorry that I did not have a good photo of Jack himself. We did correspond some, and later I sent him the special Oct. 15, 1950 edition of the *Tacoma News Tribune* on the opening of the bridge.

The last load I lifted on the new Narrows Bridge was the dampeners at the west midspan tower. In size, they made me think of a 10-gallon milk can. These shock absorbers are steel hydraulic cylinders. Placed under the deck, they were then connected to the tower at a 45-degree angle from the roadway's stiffening trusses. The concrete road would be poured next, but my job was finished. I went on to another with Haeuser and Seifert in Puyallup.

In 1984 when writing this book seemed like a good idea, I wrote to the Bethlehem personnel department giving Jack's Durkee's address of 34 years ago. My letter was forwarded to Jack in Bethlehem, Pennsylvania, and we have corresponded several times since. He now is on the International Board of Consulting Engineers, planning a bridge between Italy and Sicily, at the top of his profession without losing that "common touch." I will end this chapter by sharing his encouragement expressed in his 1989 letter.

"Yes, Joe, I remember well the many conversations we had on the boat and out on the bridge. TNB was a great experience, and you and I were fortunate to be a part of it.

"Most people have no idea how a large bridge is built. You and I have an insight into this activity, having been on the scene during that historic reconstruction. Looking at the bridge now...all of the flavor of the construction days...is replaced by the incessant rushing traffic. Who could forget the ambience of TNB under construction? Something like 100 site workers did the whole job—backed up of course by thousands more off-site. It was an enterprise to remember and report on. Your book is needed."

J. L. Durkee, field engineer for Bethlehem Steel, carries his camera on the TNB job. Joe Gotchy photo

Bridging The Narrows

Taken early on a May morning, this view of the bridge shows it still lacking its midspan. Jack Durkee photo

Bridging The Narrows

Jack Durkee Photo

During a move, Gotchy crossed space on the catwalk to turn around and take this shot of the west midspan approaching the center. Joe Gotchy photo

Bridging The Narrows

About to meet its midspan counterpart, the west midspan nears completion in this view showing the 33-foot depth of the road deck. Joe Gotchy photo

The final top chord is placed by the traveler crew. Joe Gotchy photo

Bridging The Narrows

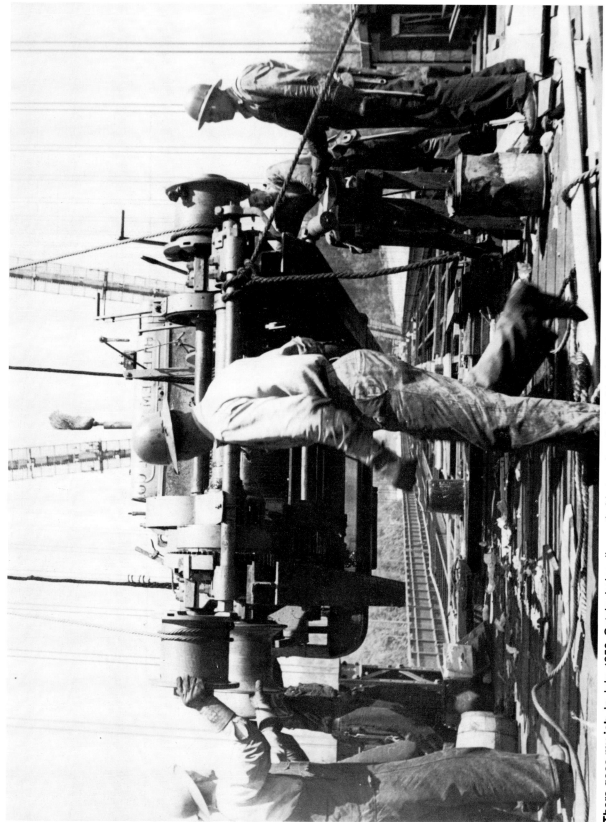

There goes my job. In July, 1950, Gotchy helps dismantle his traveler after backing it up from center to tower. With the help of the Chicago boom it will be loaded onto scows waiting below. Jack Durkee photo

Bridging The Narrows

Reinforcing rods are in place in road deck, for the concrete pours and final asphalt layer. The new superstructure weighs only 1.6 times as much as the two-lane bridge it replaces, according to the Andrew report. Pioneer, Inc. pamphlet

The steel deck is complete. This view looks east toward Tacoma. Joe Gotchy photo

Bridging The Narrows

The Tacoma Narrows Bridge on opening day, Oct. 14, 1950, showing tollbooth plaza on east approach. Photo by Harry R. Boersig, Aero-Marine Photos, courtesy Bill Loomis

Chapter Seven

A TRIP TO THE TOP OF THE EAST TOWER

On July 7, 1989, as I was leaving the Span Deli on the west end of the Narrows Bridge the proprietor, Ron Jones, stopped me and said, "Joe, let me introduce you to another bridgeman. This is Kip Wylie, supervisor of maintenance on the bridge."

When Kip found that I had worked on both bridges he told me I would be welcome on a trip to the top of the east tower. He delegated Bob Beard to conduct us on our visit. My friend, Warren Medak, who had worked on Galloping Gertie as well as on the second towers and cable spinning, was invited along. We looked forward to this rare opportunity and arrived at the shop under the deck at the east cable anchorage at 7:45 a.m. the next day.

We met the entire maintenance crew: Dennis Ulmer, supervisor; Kip Wylie, lead technician; Bob Beard, Robin Jones, Jon Moergen and David Santos, all suspension bridge maintenance men.

First we were taken back to the tunnel leading to the cable anchorage, where the cable is splayed out into individual strands and each is separately anchored. From there we went to the base of the east tower at deck level. A steel door was unlocked so we could climb down to a lower level beneath the roadway deck. Quite a few logging trucks passing over the expansion joints sure do cause a lot of vibration and movement. One can feel it as well as hear the shuddering, which tempts one to wonder what the life of such a structure truly is.

The maintenance crew live with this every day. Next, we went to the elevator going to the tower top. Anyone with even a mild case of claustrophobia should think twice before taking the elevator to the tower top. A small, circular cage, with room for only one person, hangs within a circular cell extending from bottom to top of the tower. First one must open the outer door of the cell, then the inner door of the cage and step inside. After closing the outer door and then the inner door the buttons directly in front of the passenger will activate the lift motor. There is just enough room in the cage to press those buttons to start or stop the cage and one may land at different levels. Upon landing and stepping out of the cage and its cell, the rider can

Bob Beard and Jon Moergen of the Tacoma Narrows Bridge maintenance crew, met Gotchy during a visit to the top of the bridge tower, summer, 1989. Joe Gotchy photo

The ladder seen here led Gotchy from a trapdoor in the deck down to the door of the elevator enclosed in a tower leg. This view looks crosswise of the deck and beneath the road. Camera case shows proportion. Joe Gotchy photo

Bridging The Narrows

A recent visit inside the west anchorage afforded these views of a main suspension cable's hold upon the earth. Clockwise from upper left: The south cable enters the west anchorage; Jon Moergen, with the entrance behind him, shows the beginning of splay into 19 bundles; Each bundle separates to encircle a shoe held in the concrete by a buried eyebar; Closeup view of continuous strands of main suspension cable. Joe Gotchy photos

choose to leave one door open, in which case the elevator cannot move. If someone else wishes to make use of the elevator both doors must be closed, allowing him to bring it down by another button. I knew I could not be stranded as long as I left one door open, and I found it an interesting experience.

Up and out on the tower, I was surprised to find that the cable saddle shaken off by the earthquake in 1949, which lay at the bottom of Puget Sound for three days before being retrieved, is inclined to show rust much sooner than its mate that remained in place.

The maintenance men are on speaking terms with all the problems that do or may arise, such as tower rollers that do not roll any more. I saw two such cases, in which the men had taken some of the weight off of rollers long enough to make repairs. A problem had developed with the rollers situated between the tower top and the saddle cradling the cables. Rollers are installed to equalize strain on the cables where they pass over the saddles. They were lubricated when installed, but lubricant does not last forever. The maintenance crew figured a way to drill and install alemite fittings so they now can be lubricated easily at required intervals.

Looking down from the tower top, more than 650 feet above the rushing water we had bridged, I recalled all the activity we hundred or so men had put into this structure. And I knew again how proud we and our families were, without saying so, that in our way we had been able to complete something that is a monument to the progressive spirit of Pierce County and Washington State.

Beard stands tall on the steel saddle at top of east tower. Joe Gotchy photo

APPENDIX

Some Specs on the two Narrows Bridges
Glossary of Terms
TNB Crew on Deck Steel
Letter to Joe Gotchy from Jack Hamilton
TNB Layout Sections

SOME SPECS ON THE TWO NARROWS BRIDGES

	GALLOPING GERTIE	TACOMA NARROWS BRIDGE
Total Length	5,939 ft.	5,979 ft.
Suspension bridge section	5,000 ft.	5,000 ft.
Center span	2,800 ft.	2,800 ft.
Shore suspension spans, each	1,100 ft.	1,100 ft.
East approach and anchorage	345 ft.	365 ft.
West approach and anchorage	594 ft.	614 ft.
Center span height above water	195 ft.	187.5 ft.
Width of roadway	26 ft.	49'10"
Width of sidewalks, each	5 ft.	3'10"
Diameter main suspension cable	17-1/2"	20-1/4"
Weight main suspension cable	3,817 T	5,441 T
Weight sustained by cables	11,250 T	18,160 T
Number #6 wires each cable	6,308	8,705
Weight shore anchors	52,500 T	66,000 T
Towers:		
Height above piers	425 ft.	467 ft.
Weight each tower	1,927 T	2,675 T
Piers:		
Area	118'-11"x65'-11"	119'x66'
East pier, total height	247 ft.	265 ft.
weight		65,000 T
depth of water	140 ft.*	135 ft.*
bottom penetration	105 ft.*	90 ft.*
West pier, total height	198 ft.	215 ft.
weight		52,000 T
depth of water	120 ft.	120 ft.
bottom penetration	55 ft.	55 ft.

*Note: Figures for depth of water and penetration for the east pier were made public upon completion of construction, 1940. An undated *Tacoma News Tribune* clipping in my possession gives the figures for Gertie appearing above. In his 1952 final report Chas. E. Andrew, principle consulting engineer, quotes the second figures for the present bridge.

Similarly, a discrepancy in the number of pier anchors has developed through the years. A TNT article says 24 concrete anchors were towed in and dumped over the west pier site; Charles Andrew cites 26 surrounding each pier. Clark Eldridge, bridge engineer employed by the state, is quoted in the July 4, 1940 issue of *Pacific Builder and Engineer* stating these figures: Each concrete-mass anchor weighed 550 tons; 24 were dropped on the west pier site; 32 were dropped on the east pier site, which was severely irregular and more affected by the tidal motion.

GLOSSARY

ALEMITE...A kind of grease fitting that allows grease to be pumped into a bearing.

ANCHORAGE...A reinforced concrete mass situated at each end of a suspension bridge to which are anchored the main cables.

BARGE...A towed, non-powered vessel usually with rounded ends, sometimes made by tearing down a former boat hull.

CAISSON...The first, or lower, section of a pier, pre-formed and towed to site.

CHICAGO BOOM...An long, fabricated steel beam fastened to something solidly at bottom incorporating a swiveling device. At top a pulley carries lines allowing it to be raised, lowered, or swung right or left.

CHORD...A main member of the bridge deck.

CLAMSHELL...A steel, powered shovel, that closes on its load like a clam to hold, lift and drop by cables controlled by an operator. Can be used underwater.

DAMPENERS...Hydraulic cylinders used as shock absorbers, located under the roadbed on each tower and at other points.

DECK...The foundation of the road surface of a bridge.

DOG (noun)...A curved hook used to fasten a load or line by engaging a notch in a winch drum.

DOG (verb)...The act of hooking the dog to secure a load.

FAIRLEAD...A device intended as guide for lines to prevent their fouling while moving.

HEADACHE BALL...The heavy iron piece giving weight to the line and hook on a derrick.

JEWELS...Cast zinc pieces terminating each suspender cable that hangs from the main cable. The jewels are placed under the road deck, which rests upon them and is supported by them.

LAYLEG...The fixed support of a derrick mast.

MIDSPAN...Specifically the center of the completed bridge.

PEDESTAL...Top portion of a pier, slightly above water level, supporting the tower.

Bridging The Narrows

PIER...The underwater foundation of a bridge. The Tacoma Narrows Bridge has two main piers, constructed of watertight concrete sections.

POINT...The site of a riveting crew's activity.

POCKETS...The special receptacles on a scow designed to carry removable concrete buckets.

PUSHER...Foreman of an ironworkers crew.

REEVE (verb)...To thread line or cable through a series of sheaves (pulley wheels) to accomplish a specific task of moving or hauling.

RIG (verb)...To prepare equipment with cable or rope for any job.

RIGGER...A worker well experienced in moving heavy equipment.

SADDLE...A device made of cast steel, fitted to the top of the bridge tower to receive the main suspension cable laid across it.

SCOW...A non-powered, towed vessel with flat bottom slanting upward at each end. While a barge and a scow are separate definitions, the terms often are used interchangeably.

SHEAVE (SHIV)...The grooved wheel inside the block of a block-and-tackle assembly, known to men in other occupations as a pulley; the shiv will differ in size to accommodate the line that must pass through it.

SUSPENDER CABLE...Also called hanger rope, it is connected to the road deck from the main suspension cable.

TRAVELER...Self-contained machinery and operating controls, used to lay steel decking components.

TREME (tree-mee)...An extendable tube of sheet iron, 12" to 18" diam. with a funnel top, used to conduct poured concrete underwater to seal the bottom.

WINCH...A working assembly of one or more powered drums and their lines set up for a specific task.

THE TACOMA NARROWS BRIDGE CREW on DECK STEEL

BETHLEHEM RESIDENT ENGINEER...Thomas Martinsen

BETHLEHEM FIELD ENGINEER...Jack Durkee

WEST PIER SUPERINTENDENT...Jesse Cheeley

EAST PIER SUPERINTENDENT...Jack Hamilton

WEST PIER FOREMAN...Frank Butler; Danny Lowe

EAST PIER FOREMAN...Don Caissin; Phil Orlando

DECK STEEL OPERATORS...Jim Bigelow; Dick (?); Joe Gotchy; Johnny Green; Carl Gustafson; Pewee Johnson; Junior (?); Ken (?); George Marsh; Art Olson; Bernie Palmatier; Paul Petrie; Scully Webster.

IRON WORKERS...Ed Badnooz; Slim Bonner; Glen Brown; Steve Carter; Dinny Cheeley; Coble; Pinetree Colby; L. Cox; Whitey Davis; Leonard Dean; Doyle Dunn; Stuart Gale; Johnny Gordon; T.M. Hanson; Red Henneberry; George Hicklin; Hughes; Ted Joslyn; Jim Johnson; Ed Kirkwood; Elmer L. Larwood; Lewellyn; Jimmy Mathews; Bob McKinney; Hank Meir; Hal Mousseau; Moyers; Bert Newell; A.M. Peterson; H.L. Peterson; Pinee; Harvey Pollard; Don Reardon; Serfass; Hap Shaffer; Buster Thompson; Thornberry; Ray Trout; Hank Vallee; Bill Vogel; Al Waller; Warner.

STATE INSPECTORS...Bob Evans; Bill Grant; Harry (?); Bill Hermsen; Kenny Arkins, Chief. (Kenny and I rode motorcycles together when we were kids in Olympia.)

Bethlehem Pacific Coast Steel Corporation

DISTRICT OFFICE
11100 SOUTH CENTRAL AVENUE

FABRICATED STEEL CONSTRUCTION

M. H. FRINCKE
MANAGER OF ERECTION

H. M. PITNEY
ASST. MANAGER OF ERECTION

P. O. BOX 58 WATTS STATION

Los Angeles 2, Calif.

Tacoma, Washington

July 21, 1950

To Whom it May Concern:

This letter is to certify that Joseph F. Gotchy was in our employ from March 29, 1950 to and including July 6, 1950, as a hoisting engineer.

His job was that of operating one of our four travelling derricks, which involved handling 3 drums and a swinger. This particular operation demanded precision work and also a major safety responsibility.

Mr. Gotchy, without exception, was one of our best operators , as well as being dependable, congenial and responsible.

Sincerely yours,

J. C. Hamilton
Superintendant

jch:msh